Shape Memory Alloy Actuators

Shape Memory Alloy Actuators

Design, Fabrication, and Experimental Evaluation

Mohammad H. Elahinia

Department of Mechanical, Industrial, and Manufacturing Engineering
University of Toledo, Toledo, OH, USA

This edition first published 2016
© 2016 John Wiley & Sons, Ltd.

Registered Office
John Wiley & Sons, Ltd, The Atrium, Southern Gate, Chichester, West Sussex, PO19 8SQ,
United Kingdom

For details of our global editorial offices, for customer services and for information about how
to apply for permission to reuse the copyright material in this book please see our website
at www.wiley.com.

Library of Congress Cataloging-in-Publication Data

Elahinia, Mohammad H.
  Shape memory alloy actuators : design, fabrication, and experimental evaluation / Mohammad
Elahinia, Department of Mechanical Engineering, University of Toledo, Toledo, OH, USA.
    pages cm
  Includes bibliographical references and index.
  ISBN 978-1-118-35944-0 (cloth)
1. Actuators–Materials.   2. Shape memory alloys.   I. Title.
  TJ223.A25E43 2015
  621–dc23
                                                              2015013573
A catalogue record for this book is available from the British Library.

Set in 10/12.5pt Palatino by SPi Global, Pondicherry, India

Printed in Singapore by C.O.S. Printers Pte Ltd

1   2016

Contents

List of Contributors

Hashem Ashrafiuon, Ph.D.
Professor, Department of Mechanical Engineering
Director, Center for Nonlinear Dynamics and Control
Villanova University
Villanova, PA, USA

Francesco Bucchi, Ph.D.
Research Fellow
Department of Civil and Industrial Engineering
Università di Pisa
Pisa, Italy

Mohammad H. Elahinia, Ph.D.
Professor, Mechanical, Industrial and Manufacturing
Engineering Department
Director, Dynamic and Smart Systems Laboratory
The University of Toledo
Toledo, OH, USA

Christoph Haberland, Dr.-Ing.
Siemens AG
Power and Gas Division
Berlin, Germany

Mahmoud Kadkhodaei, Ph.D.
Associate Professor
Department of Mechanical Engineering
Isfahan University of Technology
Isfahan, Iran

Haluk E. Karaca, Ph.D.
Associate Professor
Department of Mechanical Engineering
University of Kentucky
Lexington, KY, USA

Mohammad J. Mahtabi
Graduate Research Assistant
Mechanical Engineering Department
Mississippi State University
Mississippi State, MS, USA

Reza Mirzaeifar, Ph.D.
Assistant Professor
Mechanical Engineering Department
Virginia Polytechnic Institute and State University
Blacksburg, VA, USA

Soheil Saedi
Graduate Research Assistant
Mechanical Engineering Department
University of Kentucky
Lexington, KY, USA

Sayed Mohammad Saghaian
Graduate Research Assistant
Mechanical Engineering Department
University of Kentucky
Lexington, KY, USA

Nima Shamsaei, Ph.D.
Assistant Professor
Mechanical Engineering Department
Mississippi State University
Mississippi State, MS, USA

Masood Taheri Andani
Graduate Research Assistant
Department of Mechanical Engineering
Virginia Polytechnic Institute and State University
Blacksburg, VA, USA

Ali S. Turabi
Graduate Research Assistant
Mechanical Engineering Department
University of Kentucky
Lexington, KY, USA

Preface

Shape memory alloys (SMAs) have widespread uses in biomedical, aerospace, disaster mitigation, and automotive applications. Shape memory effect and superelasticity as the distinct thermomechanical behaviors of these materials have been harnessed primarily in nickel–titanium (NiTi), the most commonly used SMA. The alloy's ability to recover large deformation in a controllable manner has led to the development of minimally invasive procedures for treating various cardiovascular conditions. This has resulted in achieving better outcomes for patients while also reducing the cost of these procedures. Lighter and simpler automotive actuators made from NiTi offer better fuel economy and comfort. Planes equipped with light and compact SMA actuators have achieved enhanced aerodynamics performance in flight tests.

The purpose of this book is to provide a text for teaching both the fundamental and practical aspects of designing SMA actuators. Each chapter focuses on one of the aspects of these actuators, including modeling, device development, control, fatigue, fabrication, and experimental evaluation. Complementary information, including experimental data and a series of computer codes for modeling the behavior of SMAs, is available for download at http://smartsys.eng.utoleod.edu.

Chapter 1 provides an introductory materials science background for SMAs while providing several examples of their applications in various disciplines. This chapter also introduces a rotary SMA actuator that is used in the rest of the book. A model for this system is introduced to discuss the phase transformation kinetics of these alloys.

Chapter 2 provides an alternative framework for developing closed-form semianalytical models for SMA actuators by reducing their three-dimensional constitutive equations. There are a few issues such as convergence difficulties and parametric sensitivity when phenomenological constitutive models are numerically implemented. This approach, which is applicable to various length scales, addresses some of these problems.

Chapter 3 covers several actuation mechanisms and methodologies in the context of 10 unique automotive and biomedical applications. In these devices, both shape memory and superelastic materials are used in the form of wires, springs, and three-dimensional shapes. For each device, the behavior of the material is explained to develop the design procedure.

Chapter 4 deals with control as an integral part of most SMA systems. In addition to designing control algorithms for these actuators, the fundamental issues that should be considered in the controller design are highlighted.

The unique mechanical behavior of the SMAs is the reason for their different fatigue response, as explained in Chapter 5. In addition to structural fatigue, these alloys are affected by functional fatigue, the degradation of their shape memory, and superelastic response.

Chapter 6, in addition to providing a comprehensive review of the well-established fabrication methods for NiTi, highlights several alternative methods such as additive manufacturing for producing functional SMA components. These approaches have the potential of simplifying the production of shape memory and superelastic devices and offer the possibility of applying them in different areas such as patient-specific medical solutions and treatments.

Chapter 7 teaches the main experimental methods for characterizing the behavior of SMAs while explaining the effect of various treatment steps on the microstructure of the material, which in turn produces their distinct thermomechanical behaviors.

I hope that this book and the accompanying materials will help students learn about SMAs and facilitate the application of these materials toward innovative solutions.

MOHAMMAD H. ELAHINIA
PROFESSOR OF MECHANICAL ENGINEERING
THE UNIVERSITY OF TOLEDO
TOLEDO, OH, USA

Acknowledgments

I would like to start by expressing my gratitude to my parents Abolfath and Zahra. They are the most generous individuals that I know. When I was growing up—while we were not wealthy—I did not know of any other kid who had as many books as I had. Throughout my life, their belief in me has never faltered, and I am deeply and endlessly grateful for their love. My uncle Gholamhossein has been my role model for as long as I can remember. I am grateful to him for being the force that pushed me in the right direction at critical moments in my life. I cannot forget Dr Farshid Asl, who has been a great mentor, motivator, and friend for the past 25 years.

This project has been more than 15 years in the making. It all started when Dr Hashem Ashrafiuon, my graduate advisor at Villanova University, introduced me to control of shape memory alloy (SMA) actuators. I am fortunate to have him as the coauthor of Chapter 4 which discusses control of actuators based on these alloys. My mentor and PhD advisor at Virginia Tech, Dr Mehdi Ahmadian, has had the greatest effect on my professional life. He has taught me the importance of seeing every situation positively. He was the one who supported me while I studied the effect of dynamics on the behavior of SMA actuators. Throughout my academic and research career, I have been blessed to work with many talented students and collaborators. Ehsan Tarkesh Esfahani came to our lab with a wealth of knowledge in robotics. His research initiated our lab's work in the area of assistive and rehabilitation devices based on the use of SMAs. Majid Tabesh, Ahmadreza Eshghinejad, Walter Anderson, and Masood Taheri Andani were the force behind our work in the area of modeling and SMA application in medical devices. Some of these four students' work is the basis for Chapter 3 on design of SMA actuators. Francesco Bucchi was a visiting PhD student in our lab and contributed significantly to the same chapter. Dr Christoph Haberland is an excellent mechanical engineer with a profound understanding of materials science and engineering. He came to our lab as a postdoctoral fellow and led our efforts in additive manufacturing of these alloys. He is a coauthor of Chapter 1, the introductory chapter, as well as Chapter 6 on manufacturing. Our group has long benefited from the help and support of Dr Mahmoud Kadkhodaei and Dr Reza Mirzaeifar who are the two leading experts on understanding and modeling the behavior of SMAs. They are the coauthors of Chapters 1 and 2. Chapter 5, on fatigue, would not have been possible without Dr Nima Shamsaei, who has academic and industrial experience in the area of fatigue and fracture. Dr Haluk Karaca's experience and expertise in experimental evaluation of SMAs has tremendously benefited our group. He has been the lead coauthor of Chapter 7 which deals with the experimental evaluation of these alloys.

I am thankful to the undergraduate, graduate, and visiting research assistants in our group. While I cannot name them all, I am especially impressed by and thankful to Ted Otieno who helped with this book while handling the full load of being a successful honors student.

Last but not least, I am forever indebted to my wife Fatemeh and our daughters Hedyeh and Hoda for their love and support. They've tolerated my lengthy ramblings about SMAs over many family dinners; without them, this and many other projects would not have been possible. Hedyeh's artistic talent and attention to detail are well captured in the cover of this book.

1

Introduction

Christoph Haberland, Mahmoud Kadkhodaei
and Mohammad H. Elahinia

This chapter is on introductory materials on shape memory alloys (SMA) behavior. Shape memory effect, and superelasticity will be covered. In this context, the benefits of SMAs in actuation will be highlighted. Phase transformation as the underlying phenomenon for the unique properties of these alloys will be presented and discussed. Different actuation mechanisms and designs will be presented and compared. Example of aerospace, automotive, industrial, and biomedical applications of SMA actuation will be used to discuss the benefits and limitations of actuations using these alloys. Particular attention will be on rotary SMA actuators. This type of actuators will be used as a continuous example throughout the book.

1.1 Shape memory alloys

SMAs are distinguished from conventional metallic materials by their ability to restore their shape after large deformations, which can significantly exceed the actual elastic deformability of the material. This is referred to as

Shape Memory Alloy Actuators: Design, Fabrication, and Experimental Evaluation, First Edition. Mohammad H. Elahinia.
© 2016 John Wiley & Sons, Ltd. Published 2016 by John Wiley & Sons, Ltd.

Shape memory effect (SME) characteristic and was first observed in 1932 in a gold–cadmium[1] alloy following a thermally induced change in the crystal structure [1, 2]. Nearly 20 years later, Chang and Read [3] identified the fundamental mechanisms in the crystal lattice and attributed this phenomenon to a thermoelastic behavior of the martensitic phase. In the following years, the SME was observed in other alloys, more than 25 binary, ternary, or quaternary alloys and alloy systems are now known to show shape memory properties [4]. In contrast to nickel–titanium (NiTi), majority of these systems however have only been considered in principle and as such have not yet achieved any practical technological importance [5]. In NiTi, the SME was observed for the first time by Buehler et al. at the US Naval Ordnance Laboratory (NOL, White Oak, Maryland) in the 1960s [6, 7]. Because of the place of discovery, besides NiTi or TiNi, the term nitinol is also commonly used for this alloy. The application of SMAs spans a wide range of length scales, and these alloys are now used in multiscale devices ranging from nanoactuators used in nanoelectromechanical systems to very large devices used in civil engineering applications. SMA devices range from simple parts like cell phone antennas or eyeglass frames to complicated devices in mechanical [8–10], biomechanical [11–13], aerospace [14], and civil engineering [15].

Today, more than 90% of all commercial shape memory applications are based on binary NiTi or ternary NiTi-Cu and NiTi-Nb alloys [5]. This is despite the relatively high world market prices for high-purity nickel and especially for high-purity titanium. It should be noted that the price of Fe- or Cu-based SMAs is lower. Additionally, as explained in Chapter 6, the manufacturing processes of NiTi are complex and challenging, which adds to the production costs. The main reason for the dominance of NiTi-based SMAs is due to their excellent structural and functional properties. The SME in NiTi allows for relatively large reversible deformations of up to 8%, characterized by good functional stability [5, 16–18]. In addition, NiTi has good wear and corrosion resistance and biocompatible properties, making it an attractive candidate for various medical applications such as surgical tools, stents, or orthodontic wires [19–22]. Furthermore, the low stiffness of NiTi attracts interest for use in bone implant applications and in regenerative medicine [23]. For actuation and motion control applications, this alloy can be easily heated by passing an electrical current while offering several advantages for system miniaturization such as high power-to-mass ratio,

---

[1] The mechanism of SME in AuCd is related to aging and is considered to be different from that of NiTi, which is due to detwinning.

maintainability, reliability, and clean and silent actuation. Due to its outstanding predominant role amongst other SMAs, in this book we mainly focus on NiTi.

The fundamental reason for the unique behavior of these alloys is due to the martensitic phase transformation. Originally, this term referred to the crystallographic phase transformation, which results in rapid cooling to a specific crystallographic phase in the Fe–C structure. This is also the basic mechanism in the hardening of steels. With increasing scientific understanding of the underlying mechanisms, the term martensitic phase transformation has been extended to a variety of other alloys (e.g., Fe–Ni or Cu–Zn) and even other material systems (e.g., some specific polymers or ceramics). In general, the martensitic phase transformation is a specific type of a crystallographic phase transformation in the solid state. When cooling the material from the high-temperature phase (β-phase, austenite), the material transforms into a low-temperature phase (α-phase, martensite). This transformation is diffusionless and therefore can occur at very low temperatures. Since no diffusion processes take place, the local concentration of the chemical composition is not affected; only the crystal structure changes. Usually this change in crystal structure is driven by a shear process, which can be described by a coordinated, cooperative movement of atoms in the crystal lattice. This results in the formation of plate, lenticular, or acicular martensite crystals.

In polycrystalline materials, the martensitic transformation is typically a heterogeneous nucleation process that starts at favored nucleation sites such as phase or grain boundaries, precipitates, or crystal defects. It should be emphasized that alloys that demonstrate martensitic transformations do not necessarily have the shape memory properties. This is due to the fact that the martensitic transformation induces high mechanical stresses. The compensation of these stresses is usually associated with irreversible processes, such as dislocation slip. In alloy systems, which have shape memory properties, the martensitic transformation is to a large extent reversible. In these alloys, the transformation stresses are compensated by twinning processes and self-accommodation of favored martensite variants. Due to the fact that no additional lattice defects are created in these reversible processes, this type of martensitic transformation is also called thermoelastic martensitic phase transformation [24, 25].

Generally, the SME can occur in three related phenomena. Two of these are the thermally induced one-way and two-way shape memory (pseudoplastic) effects. The third effect is the pseudoelasticity, which is also known as superelasticity and mechanical memory. Several factors define which of these effects takes place. The most important factor is the alloy composition.

Others include the thermomechanical treatment, the microstructure of the material, and the ambient temperature [26, 27]. Generally, the binary alloy nickel–titanium is known to show all three effects depending on these factors.

Shape memory effect, shown in Figure 1.1b, as the ability of these alloys to recover a certain amount of unrecovered strain upon heating, takes place when the material is loaded such that the structure reaches the detwinned martensite phase and then unloaded while the temperature is below the austenite start temperature (A_s). Heating the material at this stage to austenite will lead to strain recovery, and the material will regain its original shape. The combined stress–strain-temperature diagram as shown in Figure 1.1b can better explain this phenomenon. During SME, as the result of cooling, the twinned variants directly transform from the austenitic phase (see Figure 1.1a, transformation $\beta \rightarrow \alpha^+/\alpha^-$). In loading the material transforms to detwinned martensite resulting in large deformation before reaching the yield stress and therefore without dislocation, beyond which dislocation plasticity starts (transformation $\alpha^+/\alpha^- \rightarrow \alpha^+$). During the heating transformation to austenite and macroscopic initial shape recovery takes place (transformation $\alpha^+ \rightarrow \beta$). To complete the cycle by cooling, the austenite transforms into self accommodating twinned martensite, without an apparent shape change (transformation $\beta \rightarrow \alpha^+/\alpha^-$).

Starting from point A, as shown in Figure 1.1b, the material is initially in the austenite phase. Cooling the alloy to a temperature below its martensite

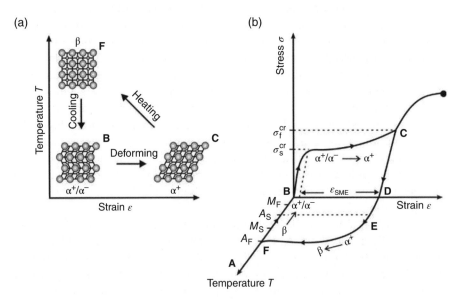

Figure 1.1 Shape memory effect path in stress–strain–temperature space

finish temperature (M_f) will result in the twinned martensite crystal, point B. This forward transformation starts when the temperature reaches martensite start (M_s). Loading the alloy at a constant temperature from point B results in elastic deformation of the martensite phase. At the critical stress levels σ_s^{cr} and σ_f^{cr}, the transformation to the detwinned martensite phase starts and finishes. This part of the loading cycle induces a large strain with minimal increase in the stress to point C. As loading continues, the detwinned martensite is elastically deformed beyond point C. Further loading will induce plastic deformation at the detwinned martensite phase.

Unloading by removing the applied stress at a constant temperature from point C results in a linear strain recovery to point D. During this step, the material remains in the detwinned martensite phase. At point D, the remaining strain is called the residual strain ϵ_{SME}. This strain can be recovered by increasing the temperature. By heating the alloy above the austenite start temperature (A_s) at point E, the transformation from the detwinned martensite to the austenite phase starts. At point F when the alloy passes the austenite finish temperature (A_f), the material is fully austenite and recovery of the residual strain completes.

In contrast to the thermally activated effects, in superelasticity, no temperature change is required. Instead, an external mechanical load is used to induce phase transformation. In this phenomenon, the detwinned and favorably oriented martensite variants directly transform from the austenitic phase (see Figure 1.2, transformation $\beta \rightarrow \alpha^+$). To complete the cycle, the material is unloaded before reaching the yield stress of the detwinned martensite beyond which dislocation plasticity starts. During the unloading, the martensite transforms back into austenite. This results in macroscopically recovering the initial shape. Due to this behavior, this phenomenon is sometimes called "rubberlike material behavior," but more common are the terms superelasticity and pseudoelasticity. It is worth noting that from the materials science point of view, this behavior actually is not elastic.

Despite the fact that no thermal activation is required to observe pseudoelasticity, it can only be activated in the temperature range $A_f < T < M_d$ where A_f refers to the austenite finish temperature (where the transformation into austenite is completed) and M_d refers to the martensite dead temperature (or martensite destruct temperature) above which no stress-induced martensite can be formed because the high-temperature phase (austenite) is stable. As Figure 1.2b shows, during loading and unloading, the material shows a hysteresis because of different stress levels for the actual transformation. It is also worth pointing out that this hysteresis corresponds well to the mechanical behavior of human tissues under a mechanical load.

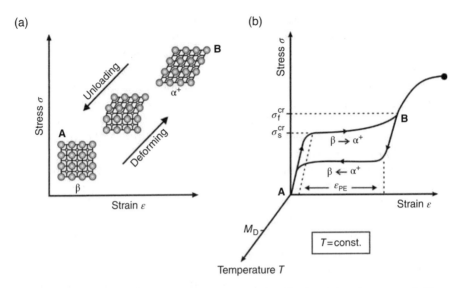

Figure 1.2 Schematic of pseudoelasticity: effect of load on crystalline structure (a); stress–strain–temperature plot (b)

The superelastic behavior takes place at temperatures above the austenite finish temperature (A_f) where the material is fully austenitic. As shown in Figure 1.2b, loading the material from point A initially induces elastic deformation of the austenite phase. Further loading leads to stress-induced formation of the detwinned martensite crystal and the macroscopically elastic-like deformation of this structure to point B. The transformation strain generated during this forward transformation from austenite to martensite is fully recovered in the reverse transformation. This takes place during the unloading from point B, which involves an initial elastic recovery of martensite followed by the transformation to austenite (recovering seemingly plastic ϵ_{PE}) and finally the elastic recovery of austenite to point A. At this point, the strain is completely recovered.

1.2 Metallurgy of NiTi

For the nickel–titanium alloy, the high-temperature phase austenite has a body-centered cubic structure (B2). This structure transforms during cooling into the monoclinic lattice structure of the martensite (B19′), whereby each lattice atom retains its nearest neighbor atoms. In binary NiTi alloys, the thermally induced SME and the pseudoelasticity only occur in a narrow range of the chemical composition. Only around the stoichiometric composition, the intermetallic phase NiTi (B2) can exist in the absence of other phases under equilibrium conditions (see gray area of the binary phase diagram shown in Figure 1.3). In addition, from the phase diagram in Figure 1.3, it is evident that the intermetallic phase NiTi has a limited solubility for titanium of less than 51 at.%. This solubility limit is almost independent of the temperature [29]. In these Ti-rich compositions and in almost equiatomic balances, the NiTi phase is stable even at low temperatures. The maximum nickel content of the NiTi phase cannot exceed 57 at.%. Below 1118°C, the solubility limit of nickel in the phase NiTi decreases with decreasing temperature. At lower temperatures, this overstoichiometric balance of the B2 phase does not exist in absence of the other phases. However, by rapid quenching from this overstoichiometric balance, the Ni-rich composition of the B2 phase can be "frozen," and the material approaches a metastable state without secondary phases.

However, if this metastable state is subjected to heat treatment, it will approach a state of thermodynamic equilibrium by complex diffusion and precipitation processes [30]. Depending on the temperature and the aging time, a two-phase state of the phase NiTi and Ni-rich precipitations will be formed. This results in a depletion of nickel in the B2 phase due to precipitation of Ni-rich phases.

Figure 1.4a shows a section of the binary phase diagram; in Figure 1.4b, the precipitation kinetics for Ni-rich NiTi ($Ni_{52}Ti_{48}$) are shown. In the beginning,

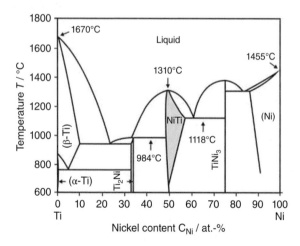

Figure 1.3 Phase diagram of the system nickel–titanium. The single-phase NiTi (B2) is shaded; important temperatures are highlighted. Reproduced with permission from Ref. [28], ASM International

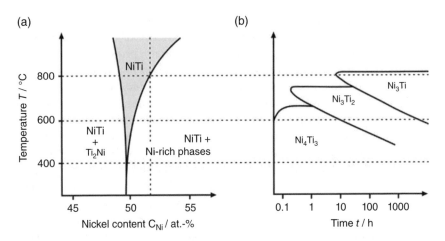

Figure 1.4   Section of the phase diagram of the system nickel–titanium (a). Isothermal transformation diagram of a $Ni_{52}Ti_{48}$ alloy (b). Reproduced with permission from Refs. [30–32], Elsevier

at low temperatures and short aging, metastable precipitates of type Ni_4Ti_3 form. With longer durations and higher temperatures, metastable Ni_3Ti_2 precipitates are formed. Precipitation of stable Ni_3Ti only occurs at very long aging treatments. The Ni_4Ti_3 phase has a significant influence on the martensitic phase transformation. With the formation of these precipitates, stress fields arise in the crystal structure, which can lead to a multistage phase transformation. In this case, during cooling from the B2 (austenite) phase, the trigonal R-phase is formed as a premartensitic state so that the transformation sequence becomes $B2 \rightarrow R \rightarrow B19'$ [33, 34]. Ni_4Ti_3 precipitates also act as nucleation sites for the martensite, and therefore, its existence reduces the critical stress to start the transformation. Moreover, the presence of the Ni_4Ti_3 particles hinders the dislocation movement, and their precipitation hardening effect leads to an increase in yield stress. As a result, the irreversible processes, which are usually associated with deformation, are reduced. Additionally, these effects contribute to an increase in the cyclic stability of the pseudoelastic effect. This in turn leads to a significant reduction in functional fatigue [35–37].

   In this metastable state, the phase transformation temperatures of Ni-rich NiTi strongly depend on the nickel–titanium ratio of the B2 matrix [33, 38, 39]. This relationship is shown in Figure 1.5 for both M_s, the martensite start temperature, and T_0, temperature of the thermodynamic equilibrium of the B2 phase and the $B19'$ phase. Binary NiTi SMAs with a substoichiometric nickel content, where NiTi and Ti-rich phases (Ti_2Ni, i.e., $Ti_4Ni_2O_X$) are in

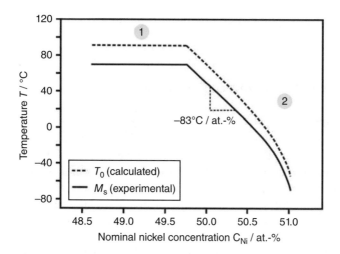

Figure 1.5 Influence of nominal nickel concentration on martensite start and thermodynamic equilibrium temperature. Reproduced with permission from Ref. [40], Elsevier

the thermodynamic equilibrium, show no significant effect (see range 1 in Figure 1.5). On the other hand, for nickel contents higher than 49.7 at.%, both temperatures continuously decrease with increasing nickel content (see range 2 in Figure 1.5). This dependence can be used to adjust the phase transformation temperatures by the nickel–titanium ratio. Alloys can thus be produced, which show either a thermal or pseudoelastic memory effect in the ambient temperature. It should however be noted that during aging of Ni-rich NiTi, the previously described precipitation processes again affect the nickel–titanium balance of the B2 matrix. During the formation of Ni-rich phases, the B2 phase is depleted of nickel, which results in an increase of the transformation temperatures [33, 39].

The effect of impurity related phases should also be considered. During high-temperature processing, the pickup of impurities, for example, carbon and/or oxygen, can result in the formation of Ti-rich phases since the B2 phase has a low solubility for both elements [41]. Carbon forms carbides of type TiC [31, 42–44], while oxygen is dissolved in the Ti_2Ni phase and forms a stable phase of type $Ti_4Ni_2O_X$ [41, 45, 46]. In addition to the degradation of functional and structural properties due to the impurity pickup, the formation of these Ti-rich phases also results in a shift of the nickel–titanium balance in favor of the nickel content. In contrast to precipitation of Ni-rich phases, this causes a decrease in transformation temperatures. These effects must therefore be given high attention in manufacturing and processing of NiTi alloys.

1.3 Thermomechanical Behaviors

The crystalline structure of SMAs undergoes a solid–solid phase transformation when cooled from its stiff, high-temperature austenite (A) phase to its softer, low-temperature martensite (M) structure. The stress–temperature–transformation (phase) plot is a schematic representation of the transformation regions for SMAs. Usually, a stress–temperature–transformation plot shows the temperature along the abscissa and stress along the ordinate. A widely accepted stress–temperature–transformation plot of SMA materials is shown in Figure 1.6 [47]. As described earlier, SMAs as the result of the transformation between two phases can exhibit SME and pseudoelasticity. The lines in the plot show the phase boundaries that separate the two solid phases of an alloy. The stress–temperature–crystalline structure pattern during shape memory effect is depicted on the

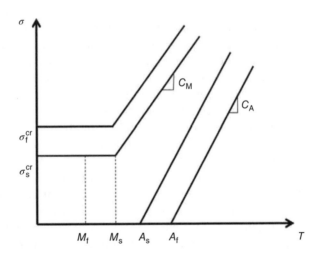

Figure 1.6 Stress–temperature–transformation (phase) plot of a shape memory material depicts the stable area for each crystalline structure. Crystal transformation takes place as the result of variation in stress and temperature (s refers to transformation start and f to transformation finish, respectively)

transformation diagram (Figure 1.7), to complement the thermomechanical and crystallin structure changes shown in Figure 1.1 for more clarity. Super-elasticity as the ability of SMAs to recover a large amount of strain through mechanical loading/unloading is shown in Figure 1.2. The stress–tempera-ture–crystalline structure cycle of the superelastic behavior is shown on the transformation plot in Figure 1.8.

As shown in Figures 1.6, 1.7, and 1.8, the four transformation temperatures that define the behavior of SMAs are stress dependent. The two parameters

$$C_A : \frac{1}{C_A} = \frac{dA_s}{d\sigma} = \frac{dA_f}{d\sigma} \text{ and } C_M : \frac{1}{C_M} = \frac{dM_s}{d\sigma} = \frac{dM_f}{d\sigma}$$

quantify the effect of stress on austenite and martensite transformation tem-peratures, respectively.

These temperatures are also affected by the thermomechanical history of the alloys. In single crystals and polycrystals of CuAlZnMn, for example, the reverse martensitic transformation is influenced by the history of the forward transformation [48]. The transformation path from austenite to mar-tensite, the deformation in martensite, and the annealing of martensite under stress affect the transformation temperatures. The incomplete and complete

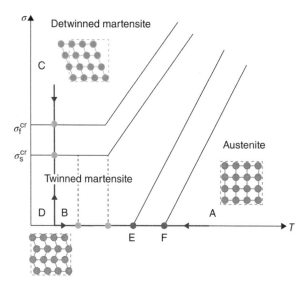

Figure 1.7 Shape memory effect and the associated crystalline changes presented in a stress–temperature–crystalline structure transformation diagram

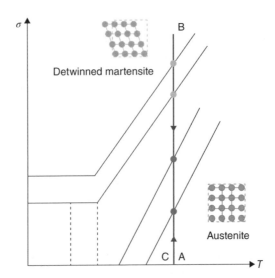

Figure 1.8 Superelasticity and the associated crystalline changes presented in a stress–temperature–crystalline structure transformation diagram

transformation cycling, that is, thermal cycling between the forward and the reverse transformation finish temperatures, in NiTi leads to variations of the transformation temperatures. Referred to as the "thermal arrest memory effect," the alloy remembers the arrested temperature in incomplete heating–cooling cycles by changing its transformation temperature in the consequent heating–cooling cycles [49–51]. When a reverse transformation is stopped (arrested) between A_s and A_f followed by cooling below M_f or in other words a complete martensitic transformation after an interrupted heating cycle, a memory of the arrest is induced in the alloy [52–55]. In a consequent heating, this memory of the arrested temperature affects the thermomechanical behavior of the alloy at an intermediate point where the transformation shows an apparent completion. If after arresting between A_s and A_f during reverse transformation the temperature is reduced below A_s but above M_s, the subsequent reverse transformation in the next heating cycle does not start at A_s. Instead, the transformation starts at the arrested temperature as shown in Figure 1.9 [56]. Similarly, for any interruption during the forward transformation, change in the martensite start temperature is observed if the specimen is subjected to heating to a temperature below A_f followed by a cooling below M_s. Moreover, if the latter heating cycle continues to a temperature above A_s, the reverse transformation does not start at A_s but at a greater temperature.

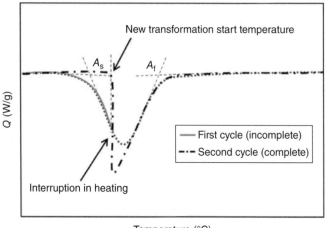

Temperature (°C)

Figure 1.9 Shifts in A_s of NiTi after an incomplete cooling cycle. Reproduced with permission from Ref. [56], Elsevier

1.4 Actuation

Recovering large strains as a result of heating is the actuation principle of SMAs. These alloys have a very high energy density; therefore, actuators that implement these alloys are compact and lightweight alternatives to other types of actuators such as DC motors and solenoids. In other words, SMA actuation is an effective way to reduce weight and to minimize the complexity of systems. The benefits of these alloys over other smart materials such as piezoelectric materials, electrostrictive materials, and magnetostrictive materials include the high force-to-weight ratio and large displacement capability. The disadvantages are slow actuation cycle due to longer cooling time, low energy efficiency due to conversion of heat to mechanical energy, and challenging motion control due to hysteresis, nonlinearities, parameter uncertainties, and difficulties in measuring state variables such as temperature.

Large mechanical stress is produced when SMA elements are heated beyond the austenite start temperature. NiTi actuators, for example, can apply up to 600 MPa stress. As a result, a 0.1 mm diameter NiTi wire can apply a force of 18.8 N, which is enough to lift 100 000 times its own weight. SMA actuators therefore offer the potential of significantly reducing the weight of active structures.

An essential step in creating the shape memory actuation is shape setting. After the material is produced and is exposed to cold work and annealing, the desired memorized shape of the actuator is instilled during shape setting. For this step, the cold-worked material is constrained in a mandrel and heat-treated with a specific duration and a predetermined temperature, which is followed by a rapid cooling. The duration is in the order of about 10 min and the temperature is around 500°C. This process is essential for both shape memory actuators as well as superelastic devices. In addition to creating the desired shape, this final heat treatment defines mechanical properties of the part. A higher temperature results in a lower tensile strength. The shape setting for actuators can also be performed continuously as a part of the fabrication process for wires and tubes, which usually result in straight memorized forms. In most cases, however, the required heat treatment for an actuator with a more complex memorized form is a separate procedure. Heat treatments and shape setting are described in detail in chapter 6.

One-way SME refers to the ability of an SMA that is deformed at a low temperature to recover the deformation when heated to a higher temperature. To create a repeatable actuation, an external bias force is used. This force always opposes the SMA actuator and during cooling resets the actuator for the next cycle. In the less common two-way SMA actuators, the SMA elements can exhibit repeatable shape changes without a bias mechanical load. Instead, the SMA alternates between memorized shapes when subjected to a cyclic thermal load. In this case, the SMA memorizes a martensite shape that is different from the austenite shape. This behavior is called the two-way SME. In this mode, when the SMA is cooled from austenite to martensite, instead of adapting to a self-accommodated structure, some variants of the martensite are favored, and the martensite adopts a shape different from that of the self-accommodated structure. This memory is usually the result of training in the form of cyclic thermomechanical loadings or aging for precipitation under stress and/or under constraint. In addition to traditional heat treatment methods, laser scanning has been used for creating the two-way memory effect [57]. It is important to note that two-way SME provide very limited strains and therefore, it cannot be used in most applications.

Multiple SME is a well-known phenomenon in shape memory polymers (SMPs) in which a deformed SMP can recover from a temporary shape back to the original shape through a number of intermediate shapes in a step-by-step manner. While this behavior is ordinarily not observed in SMAs, it is possible to induce multiple SME in these alloys [58]. To this end, a laser beam is used to locally heat-treat NiTi. This treatment causes precipitation and local evaporation of nickel, which results in different local A_f, which is higher than that of the base material. By adjusting the duration and number of pulses of laser, it is possible to increase the transformation temperatures. If this NiTi element is deformed at low temperature ($<A_s$) and consequently heated, initially, the untreated sections of the alloy transform to austenite followed by the heat-treated sections, resulting in intermediate memorized configurations. Alternatively, a series of local cooling–loading–cooling cycles can be used to induce multiple SME [59].

SMAs have been studied and deployed to a wide range of actuation mechanisms. To this end, various forms of SMAs have been fabricated from mostly NiTi and NiTiCu alloys. In terms of actuation, the main difference between these two alloys is in a smaller temperature hysteresis in NiTiCu. Wires are the most common form for actuation. Other forms that have been used are springs, thin films, sheets, tubes, rods, and more recently three-dimensional fabricated shapes.

Thin film actuators offer miniaturization and high bandwidth. The higher actuation speed is achieved due to much faster convection cooling in films. NiTi films are produced by MEMS fabrication methods in vacuum deposition, electron irradiation, and magnetron sputtering [60]. The applications include microactuation, microswitching, and micropumping [61]. Similarly, sheets are formed to 3D actuators for folding origami, microactuation, and robotic applications [62, 63]. To this end, laser cutting and shape setting are used to achieve the final shape and functionality. It should be noted that methods for fabricating NiTi are described in detail in chapter 6. Figures 1.10 and 1.11 show a sheet-based NiTi actuator that move a series of cylinder-and-sleeve joints, which allow the elements to rotate relative to one another during actuation [64, 65]. The SMA sheets are prestrained at a low temperature ($T < A_s$) to approximately one-half

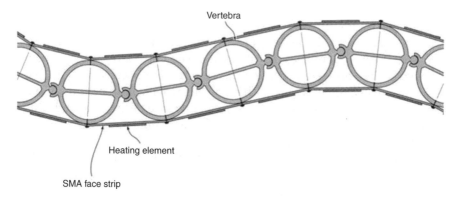

Figure 1.10 Schematic of the morphing wing structure. Reproduced with permission from Ref. [64], SPIE

(a)

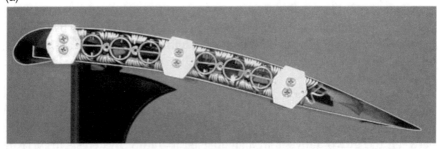

(b)

Figure 1.11 Two configurations of the shape morphing structure when (a) both actuator ribs are at minimal curvature and (b) both actuator ribs are at maximal curvature. Reproduced with permission from Ref. [64], SPIE

of their maximum recoverable strain. Heating an SMA sheet causes a contraction that results in bending of the structure. During this process, the opposite sheet undergoes a tensile strain. The curvature of the structure is reversed by consequent heating of the opposing SMA sheet. The SMA elements are heated by resistance heating elements wound helically around each sheet. Heating all the SMA sheets on one side of the structure results in a uniform curvature. Individual SMA actuators may be selectively heated to achieve various shapes.

Rod SMA actuators have been developed for breaking concrete and boulder [66]. In this actuator, a series of mechanically parallel rods, upon heat activation, push two steel wedges apart by generating up to 900 kN of axial force. Structural control applications such as morphing aerospace systems require higher level of force and torque. To achieve this, a larger cross-sectional area of the actuator is necessary. Beam actuators have been used to modify the outlet geometry of jet engines [67]. These beams are shape set to a curved geometry. By activating these NiTi beams, the attached elastic laminated structure bends causing change in the engine outlet area. This reduced engine outlet area leads to creating a mixture of jet engine fan flow with core flow. The result is a noise reduction in takeoff and landing. The outlet returns to normal for cruise flight regime.

As another aerospace application, Boeing, NASA, AFRL, and DARPA have developed NiTi tube rotary actuators also called "torque tubes". Through cyclic thermomechanical loadings, these SMA tubes are trained as a heat-activated actuator with actuation torque of $17\,\mathrm{N\cdot m}$ and angular displacement of $60°$, with repeated actuation of over $10\,000$ cycles [68]. Such an actuator can be trained for two-way actuation and is envisioned for deploying and retracting tabs on rotor blades for reducing noise and for morphing blades [69]. In 2013, an SMA rotary actuator was integrated into the hinge line of a small flap on the trailing edge of a Boeing 737–800 wing and tested in flight in a collaboration between Boeing and the FAA [70, 71]. As Figure 1.12 shows, the actuator is made of two tubes and provides $45°$ of bidirectional trailing edge flap motion under realistic aerodynamic loads. This type of actuator can be placed at various locations on the wing. Rotary actuation of the edge flap enhances the performance in various flight regimes by improving lift and optimizing span loading. At small angles, better fuel economy and lower emissions are possible during high-speed cruise. On the other hand, with larger angles, lift is increased and noise is decreased during takeoff and approach.

Wires are the most common form of SMA actuators. Joule heating is mostly used as an effective and simple way for actuating SMA wires. Miga Motor

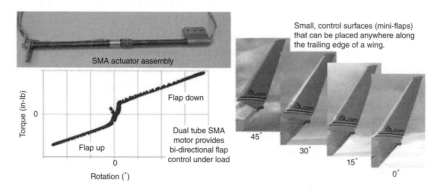

Figure 1.12 Boeing's small trailing edge flap with a rotary SMA actuator; motion data under realistic aerodynamic loads and flight tests. From Refs. [70, 71]

Figure 1.13 A linear bias-type cascaded (left) and a rotary (right) shape memory alloy actuator (NanoMuscle). Courtesy of Miga Motor Company

Company offers commercial wire-based linear and rotary actuators with integrated sensing and an on–off controller. As shown in Figure 1.13, to achieve a higher level of linear stroke, wire actuators are arranged in a cascade form. By using various length and diameter of wires, these actuators deliver a range of motion, force, and speed. A lifetime of over one million cycles at a cyclic actuation time of less than a second is achieved.

Standardization and consistent performance are needed for SMA actuators to find widespread acceptance in automotive applications. Figure 1.14 shows one such attempt as an SMA actuator which is optimized for automotive applications with long fatigue life. Using a superelastic antagonistic wire, an adaptive resetting mechanism is integrated in the actuator. This mechanism offers close-to-constant bias force which in turn allows for a larger useful stroke. The other benefit of this bias mechanism is that in reaction to the change of the ambient temperature, the superelastic wire adjusts its resistive force. When used for fuel door unlocking, this actuator reduced the weight of the actuation system by more than 90% [72].

One part of the actuation market is in individualized actuation and for replacing conventional actuators in current systems. In an attempt to address this need, as Figure 1.15 shows, an SMA spring actuator is designed with a 3D-printed polymer housing to create a thermal control valve actuator [73]. During the printing process, the seal elements, resetting (bias) springs, and SMA actuator springs are intended for automatic insertion by the handling systems. The packing material for the SMA and bias spring is ABS-based

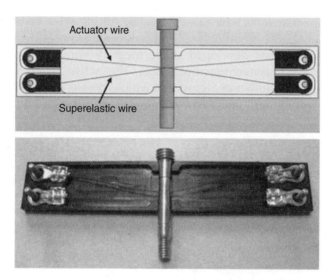

Figure 1.14 Standardized SMA actuator with adaptive resetting. From Ref. [72]

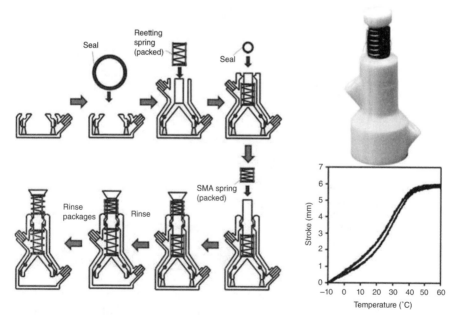

Figure 1.15 Production steps of an SMA valve actuator by 3D-printed housing. From Ref. [73]

material in the wire form which is removed by rinsing at the end of the printing process.

General Motors has developed shape memory wire-based actuators to
replace the heavier conventional actuators. The Active Hatch Vent, as one
of these devices, is shown in Figure 1.16 which consists of a U-shaped SMA
wire, a bias spring, and an assembly to create the rotary motion necessary

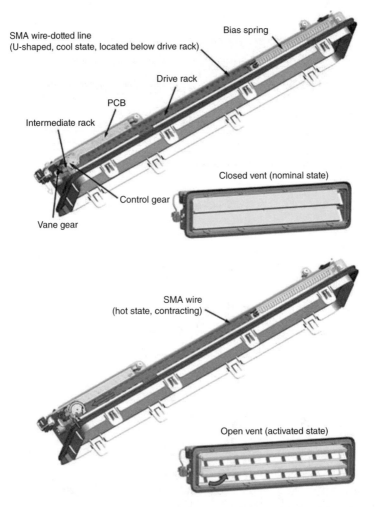

Figure 1.16 A wire-based shape memory alloy actuator developed by
General Motors to open a vent when the wire is heated. Courtesy of General
Motors, Warren, Michigan

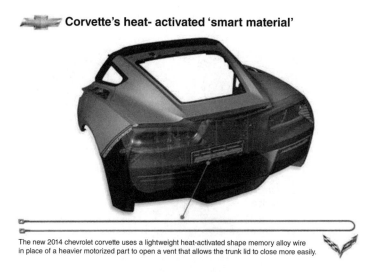

Corvette's heat- activated 'smart material'

The new 2014 chevrolet corvette uses a lightweight heat-activated shape memory alloy wire in place of a heavier motorized part to open a vent that allows the trunk lid to close more easily.

Figure 1.17 The shape memory alloy actuator shown in Figure 1.16 is applied to develop the Active Hatch Vent to open the vent in the trunk of a luxury sedan vehicle. Courtesy of General Motors, Warren, Michigan

to open a vent. This vent actuator is incorporated in the 2014 Chevrolet Corvette as shown in Figure 1.17 and has reduced the weight of the conventional actuation systems by 0.6 kg.

Maximum recoverable strain defines the stroke of SMA actuators when the actuator applies no force. There is an inherent compromise between the actuation force or torque and the deliverable linear or rotary motion: the larger the actuation force, the smaller the maximum stroke. It is however possible to overcome the limitation in stroke, as well as the compromise between the output torque and angular displacement of SMA actuators through the use of wire-on-drum modular actuators combined with unidirectional coupling mechanisms as shown in Figure 1.18 [74, 75]. The resetting bias force is provided by a beam spring which generates a nearly constant force tangential to the drum. The SMA wire is activated through Joule heating which results in the rotation of the drum. The rotation of the drum is transferred through an overrunning clutch in one direction. When the drum is rotated backward, the SMA wire is recoiled to prepare for the next step of the actuation, while the output shaft does not rotate. A second overrunning clutch links the shaft to the frame to stop accidental unwanted rotations of the output shaft.

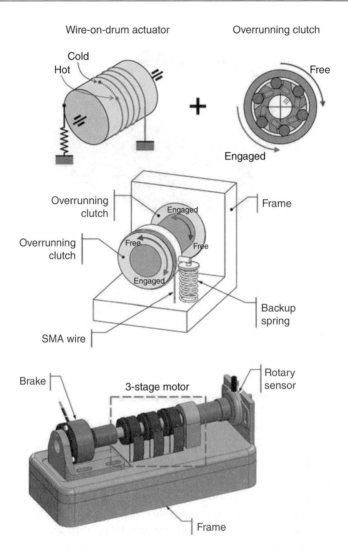

Figure 1.18 (Top) Schematic of a wire-on-drum SMA rotary actuator combined with an overrunning clutch for creating a module with unlimited angular motion; (middle) the modular combined actuator; (bottom) the modules can be combined to increase the output torque or speed in series or parallel arrangement [74, 75]. Courtesy of Dragoni and Scirè Mammano, University of Modena and Reggio Emilia, Italy

(a)

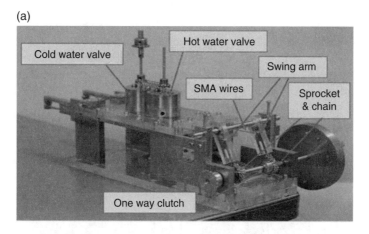

(b)

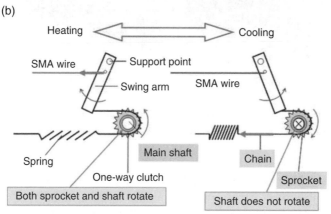

Figure 1.19 (a) Reciprocated heat engine employing an SMA wire. (b) Action mechanism of the heat engine. Reproduced with permission from Ref. [76], the Japan Society of Mechanical Engineers

In a similar use of rectifying mechanisms, one of the early applications of NiTi wire actuators was in heat engines to harness energy from sources of heat, such as internal combustion engines. An example of this type of actuators is shown in Figure 1.19 [76]. This reciprocating ratchet-type heat engine uses an SMA wire, which is actuated by alternating flow of hot/cold water through automatic valves and thereby harvesting energy from heat. A spring provides the bias force for repeated actuation. The SMA wire is connected to a chain and sprocket, which only transmits the rotation in one direction and therefore creates a one-directional rotary motion at the output shaft.

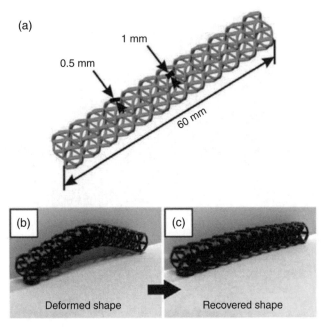

Figure 1.20 A shape memory porous beam produced in additive manufacturing using powder-bed selective laser melting [84]

Wire-based NiTi actuators have been developed for robotic [77] and positioning applications. An example is an automotive side mirror actuator [78]. Wire actuators are also embedded in polymers to create composite actuators in the form of beams or plates [79]. The off-center wires are installed on or embedded in both sides of the structure to create bending moment in the structure. The elastic substrates store the strain energy, which is released for repeated actuation. This type of actuators has been designed for biomimetic fins, morphing wings, flapping wings, stiffness variation, adaptive structures, and minirobotic applications [80–82].

There has long been an interest in three-dimensional SMA actuator elements including monolithic and porous structures [83]. Figure 1.20 depicts an SMA porous actuator that is additively manufactured using selective laser melting. This one-way Ti-rich NiTi actuator functions without the need to shape setting; the memorized configuration is the printing shape. In addition to the simplicity of manufacturing, the porous structure provides a very lightweight actuator with adjustable stiffness [84].

1.5 Modeling and Simulation

Conducting experiments on SMAs is expensive and time consuming. There-fore, in the design of SMA actuators, it is essential to have reliable modeling platforms to avoid unnecessary trials and errors. There have been many attempts to mathematically model the SMA features over the past three dec-ades. The resulting models can be divided into two categories: micromodels and phenomenological macromodels. In general, micromechanical-based models utilize information about the microstructure of the SMA to predict the macroscopic response. Micromechanical models are useful in understand-ing the fundamental phenomena of SMA, although they may not be easily deployed for engineering applications. On the other hand, phenomenological models use the principles of continuum thermodynamics to describe the material response. They are calibrated by a limited number of parameters measured at the macroscopic scale through experimental observations, and thus, they are more computationally efficient.

In a general case, the behavior of an SMA-based device could be demon-strated through the following interconnected submodels: constitutive model, phase transformation kinetics, heat transfer model, and kinematics and dynamics of the device (Figure 1.21). In this section, these submodels are dis-cussed briefly for a representative rotary actuator. Chapter 2 presents a more fundamental and comprehensive treatment of the modeling.

A one-degree-of-freedom rotary SMA actuator is shown in Figure 1.22. This system is used as an example throughout the book. This SMA system, despite its simplicity, presents several interesting and unique properties,

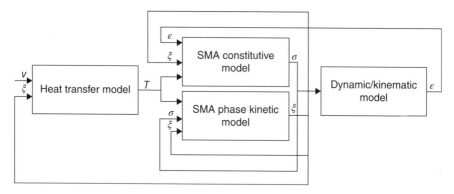

Figure 1.21 Modeling of the SMA systems; four submodels are interconnected. From Ref. [85]

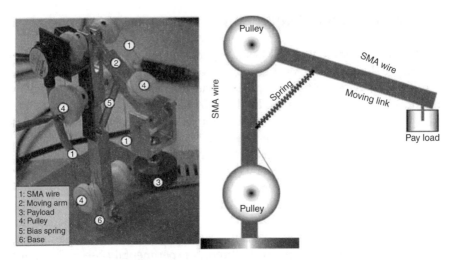

Figure 1.22 A one-degree-of-freedom rotary SMA NiTi actuator; the upward motion is generated by shape memory effect and the downward motion by the bias spring and the weight of the payload. From Ref. [77]

which make it fitting as an interesting example. To allow a more comprehensive learning experience, a complete model of the system has been made available at http://smartsys.eng.utoledo.edu. The accompanying MATLAB/Simulink code allows for simulation of the system behavior in both open- and closed-loop forms. This system and the associated model have been the subject of in a series of publications in the area of modeling and control of SMA actuators [77, 86–93].

The design parameters of the rotary SMA actuator consist of the length and diameter of the NiTi wire. The length of the wire is defined through the kinematic relationship between the angular rotation and strain of the wire:

$$\varepsilon = -\frac{2r\theta}{l_0} \tag{1.1}$$

Maximum strain takes place at the lower position of $-45°$ where the wire is assumed to be mostly in the detwinned martensite phase. At $+90°$ angular position, the actuator wire is assumed to be mostly in the austenite phase and at the minimum strain state. As shown in Chapter 5, in order to maximize the fatigue life of the SMA wire, it is recommended to limit the maximum

Table 1.1 Parameters of the one-degree-of-freedom SMA rotary actuator

Parameters	Description	Unit	Value
M	SMA wire's mass per unit length	kg	$1.414e^{-4}$
A	SMA wire's circumferential area per unit length	m	$4.712e^{-4}$
C	Specific heat of wire	kcal/kg · °C	0.2
R	SMA wire's resistance per unit length	Ω	45
T_∞	Ambient temperature	°C	20
H	Heat convection coefficient	$J/(m^2 \cdot °C \cdot s)$	150
E_A	Austenite Young's modulus	GPa	75.0
E_M	Martensite Young's modulus	GPa	28.0
θ_T	SMA wire's thermal expansion factor	MPa/°C	0.55
Ω	Phase transformation contribution factor	GPa	−1.12
σ_0	SMA wire's initial stress	MPa	75.0
ε_0	SMA wire's initial strain	—	0.04
T_0	SMA wire's initial temperature	°C	20
ξ_0	SMA wire's initial martensite fraction	—	1.0
A_s	Austenite start temperature	°C	68
A_f	Austenite final temperature	°C	78
M_s	Martensite start temperature	°C	52
M_f	Martensite final temperature	°C	42
C_A	Effect of stress on austenite temperatures	MPa/°C	10.3
C_M	Effect of stress on martensite temperatures	MPa/°C	10.3
L_0	Initial length of SMA wire	mm	900
R	Pulley diameter	mm	8.25
m_p	Payload mass	gr	57.19
m_a	Moving link mass	gr	18.7
K	Bias spring stiffness	N/m	3.871

Source: From Ref. [77].

strain to 4%. Using this strain and Equation 1.1, the initial (short) length of the wire can be calculated, as shown in Table 1.1.

In Chapter 3, it will be shown that the stress of the SMA wire in this actuator varies as a function of the angular position of the arm. This nonlinear behavior is because the torques generated by the bias spring and the payload are non-linear functions of the angular position. Additionally, the relationship between the stress of the wire and angular position of the arm is modified due to angular acceleration as captured in the dynamic model:

$$I\ddot{\theta} + c\dot{\theta} + \tau_{spring} + \tau_{load} = \tau_{SMA} \tag{1.2}$$

Using Equation 1.2, the maximum desirable actuation torque to overcome the maximum resistance torque of the spring and the payload and to achieve the desired angular acceleration can be calculated. This actuation torque is then used to calculate the diameter of the SMA wire, as shown in Table 1.1. The arm rotates from −45° to +90° when the NiTi wire with the calculated diameter of 150 μm is actuated by Joule heating.

While Chapter 2 includes more details on SMA modeling, in this section, a phenomenological model is presented to capture the behavior of the rotary actuator [90, 91]. This approach follows the assumption [94] that the material behavior of the alloys depends on the state variables strain (ε), temperature (T), and crystallographic phase (martensite fraction $\xi = 1$ when the alloy is 100% martensite and $\xi = 0$ when the alloy is 100% austenite), all of which are mutually dependent on each other as schematically shown in Figure 1.21. The parameters associated with this modeling approach are summarized in Table 1.1. Based on the second law of thermodynamics expressed in the Clausius–Duhem inequality form and expressing the thermodynamic potential as Helmholtz free energy, a constitutive equation captures the thermomechanical behavior of the SMA wire [95]:

$$\sigma = \sigma(\varepsilon, T, \xi)$$

$$d\sigma = E d\varepsilon + \theta_T dT + \Omega d\xi \tag{1.3}$$

In Equation 1.3 which was proposed and modified by Tanaka [94], Liang and Rogers [96], and Brinson [95], E, θ_T, and $\Omega = -\varepsilon_L E$ represent Young's modulus, thermal expansion coefficient, and phase transformation contribution factor, respectively. ε_L is the maximum recoverable (transformation) strain of the NiTi wire which is by design limited to 4% when the arm is at −45°. For simplicity, these parameters are assumed here to be independent of the state variables.

The second part of the model represents the state of the material by calculating the martensite fraction as a function of the temperature, stress, and thermomechanical loading history. Such a model can be found by curve fitting so that the combined phase transformation–constitutive model captures macroscopic hysteretic behavior of the material. Table 1.2 summarizes the evolution of conditions for phase transformation.

For the rotary actuator, the SMA wire, in heating or mechanical unloading, transforms from detwinned martensite to austenite. In cooling, the transformation is back to detwinned martensite. The wire does not transfrom to twinned martensite due to the presence of the bias torque. A phase transformation model in general includes the forward and reverse transformations as

Table 1.2 The evolution of phenomenological phase transformation laws

Author	Martensite to austenite transformation	Austenite to martensite transformation
Tanaka [94]	$\dot{\sigma} < 0$ $\sigma \le -(T - A_s) A_a / B_a$	$\dot{\sigma} > 0$ $\sigma \ge (T - M_s) A_m / B_m$
Liang and Rogers [96]	$\dot{T} > 0$ $A_f + \dfrac{\sigma}{C_A} \ge T \ge A_s + \sigma/C_A$	$\dot{T} < 0$ $M_s + \dfrac{\sigma}{C_M} \ge T \ge M_f + \sigma/C_M$
Brinson [95]	$\dot{\sigma} < 0$ $C_A(T - A_s) \ge \sigma \ge C_A(T - A_f)$	$\dot{\sigma} > 0$ $C_M(T - M_f) \ge \sigma \ge C_M(T - M_s)$
Elahinia [85]	$\dot{T} - \dfrac{\dot{\sigma}}{C_A} > 0$ $A_f + \sigma/C_A \ge T \ge A_s + \sigma/C_A$	$\dot{T} - \dfrac{\dot{\sigma}}{C_M} < 0$ $M_s + \sigma/C_M \ge T \ge M_f + \sigma/C_M$

Source: From Ref. [91].

well as the detwinning effect. The associate equation to capture the reverse transformation can be written as

$$\xi = \frac{\xi_M}{2} \left[\cos[a_A (T - A_s) + b_A \sigma] + 1 \right] \tag{1.4}$$

The forward transformation takes place when the wire is cooled or mechanically loaded. Variation of the martensite phase can be modeled with the following equation:

$$\xi = \frac{1 - \xi_A}{2} \cos[a_M (T - M_f) + b_M \sigma] + \frac{1 + \xi_A}{2} \tag{1.5}$$

where a_A, a_M, b_M, and b_A are the curve fitting parameters, ξ_M is the minimum martensite fraction reached during heating or unloading and ξ_A is the maximum martensite fraction reached during colling or loading.

In the more common bias type SMA actuation the material transforms to detwinned martensite during cooling. In certain SMA actuators, unlike the rotary actuation of this example, there is a possibility of transforming to twinned martensite when the actuation element is cooled under M_s at low level of stress below σ_s^{cr}. This element, when consequently heated, transforms to austenite. As an example, this phenomenon takes place, as shown in Figure 1.23, when two SMA wires form an antagonistic actuation–bias mechanism [97]. Figure 1.24 shows a similar antagonistic microactuator consisting of a bistable curved beam, two SMA active elements, and laser sources [98]. The laser beam is a contactless energy transfer source to activate the SMA elements. In each direction, one SMA active element actuates the bistable curved

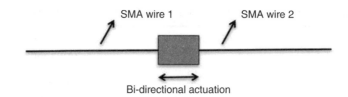

Figure 1.23   An antagonistic SMA actuator consists of two SMA elements

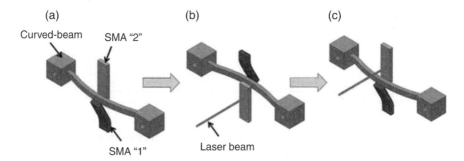

Figure 1.24   (a) Initial arrangement of the components of the contactless bistable actuator, curved beam at its first stable position, (b) curved beam switched toward the second stable position after irradiating with the laser beam, and (c) curved beam switched back to the first stable position. Reproduced with permission from Ref. [98], IOP

beam between its two stable positions. The memorized shape of each of the two SMA elements is the predetermined flat shape, which is achieved upon heating (Figure 1.24b). During the actuation, the passive antagonistic SMA element is also deformed and transformed to stress-induced martensite to prepare for the next step of actuation. The reverse switching process is obtained by heating the second SMA active element as shown in Figure 1.24c. At the end of each actuation cycle, when each active element is cooled, it transforms to twinned martensite. Other examples of such actuators as antagonistic shape memory–superelastic systems are highlighted in Chapter 3.

To capture the behavior of antagonistic SMA actuators, Brinson divided martensite volume fraction into two portions: temperature induced (twinned) and stress induced (detwinned) [95]. These fractions together form the total martensite fraction

$$\xi = \xi_T + \xi_S \tag{1.6}$$

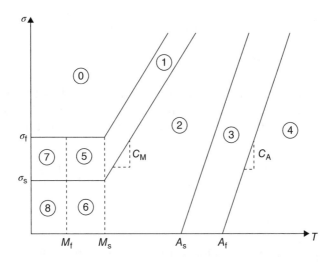

Figure 1.25 Stress–temperature $(\sigma - T)$ plane divided in nine regions to capture various transformation and detwinning possibilities

and the constitutive equation is simplified as

$$\sigma = E\varepsilon + \Omega\xi_S = E(\varepsilon - \varepsilon_L\xi_S) \qquad (1.7)$$

To capture various actuation possibilities, a phase transformation plot can be divided into 9 regions as shown in figure 1.25. In the system of Figure 1.23, before the actuation and when the actuator is in the middle neutral position, the two wires are assumed to be 50% temperature-induced and 50% stress-induced martensite. Upon complete actuation of each of the two wires, the actuating wire is transformed to austenite (region 4 in Figure 1.25), while the opposing wire is transformed to 100% stress-induced martensite (region 0 in Figure 1.25). In this antagonistic actuator, when the active wire is cooled at the end of actuation, it transforms to twinned martensite (region 8 in Figure 1.25) as long as the following two conditions are simultaneously satisfied:

$$\dot{T} - \frac{\dot{\sigma}}{C_M} < 0 \text{ and } M_f + \frac{\sigma}{C_M} < T < M_s + \frac{\sigma}{C_M} \qquad (1.8)$$

The transformation of the stress- and temperature-induced martensite variants can be described as

$$\xi_T = \frac{1-\xi_{AT}}{2}\cos[a_M(T-M_f) + b_M\sigma] + \frac{1+\xi_{AT}}{2}$$

$$\xi_S = 0 \qquad\qquad\qquad (1.9)$$

where ξ_{AT} is the twinned martensite fraction prior to this transformation. When this wire is consequently stretched (toward region 0 in Figure 1.25) by actuation of the antagonistic wire, the following opposing transformations take place:

$$\xi_S = \frac{1-\xi_{AS}}{2}\cos\left[\frac{\pi}{\sigma_s-\sigma_f}[\sigma-\sigma_f-C_M(T-M_s)]\right] + \frac{1+\xi_{AS}}{2}$$

$$\dot{\xi}_T = -\dot{\xi}_S \qquad\qquad\qquad (1.10)$$

when the following four conditions are simultaneously satisfied:

$$\dot{\sigma} > 0, \ \sigma > \sigma_s^{cr}, \ \sigma < \sigma_f^{cr}, \ \text{and} \ M_f + \frac{\sigma}{C_M} > T$$

In Equation 1.10, ξ_{AS} is the stress-induced martensite fraction prior to this transformation.

To capture other possible behaviors of the SMA element, this section summarizes the behavior in all nine zones as depicted in Figure 1.25 [99]. Within each region, the stress-induced martensite ξ_S and the temperature-induced martensite ξ_T are defined on the basis of the tensile stress σ and the temperature T, in addition to the initial stress-induced martensite ξ_{S0} and the initial temperature-induced martensite ξ_{T0}. It is worth noting that the values ξ_{S0} and ξ_{T0} have to be updated every time the boundary between two regions is crossed. Depending on the thermomechanical loading history, it is possible that the material may exist in a region with or without a phase transformation.

Region 0: Pure stress-induced martensite

$$\xi_S = 1 \qquad\qquad\qquad (1.11)$$

$$\xi_T = 0 \qquad\qquad\qquad (1.12)$$

Region 1: Transformation from austenite or from twinned martensite to detwinned martensite

$$\xi_S = \frac{1-\xi_{S0}}{2}\cos\left[\frac{\pi}{\sigma_s^{cr}-\sigma_f^{cr}}\left(\sigma-\sigma_f^{cr}-C_M(T-M_s)\right)\right] + \frac{1+\xi_{S0}}{2} \qquad (1.13)$$

$$\xi_T = \xi_{T0} - \frac{\xi_{T0}}{1-\xi_{S0}}(\xi_S - \xi_{S0}) \tag{1.14}$$

Region 2: Mixture of twinned and detwinned martensite and austenite, depending on the loading/temperature history; no transformation occurs in this area

$$\xi_S = \xi_{S0} \tag{1.15}$$

$$\xi_T = \xi_{T0} \tag{1.16}$$

Region 3: Transformation from martensite to austenite

$$\xi = \frac{\xi_0}{2}\left[\cos\left(a_A\left(T-A_s-\frac{\sigma}{C_A}\right)\right)+1\right] \tag{1.17}$$

$$\xi_S = \xi_{S0} - \frac{\xi_{S0}}{\xi_0}(\xi_0 - \xi) \tag{1.18}$$

$$\xi_T = \xi_{T0} - \frac{\xi_{T0}}{\xi_0}(\xi_0 - \xi) \tag{1.19}$$

Region 4: Pure austenite, no transformation

$$\xi_S = 0 \tag{1.20}$$

$$\xi_T = 0 \tag{1.21}$$

Region 5: Transformation from austenite or twinned martensite to detwinned martensite and transformation from austenite to twinned martensite

$$\xi_S = \frac{1-\xi_{S0}}{2}\cos\left[\frac{\pi}{\sigma_s^{cr}-\sigma_f^{cr}}\left(\sigma-\sigma_f^{cr}\right)\right]+\frac{1+\xi_{S0}}{2} \tag{1.22}$$

$$\xi_T = \Delta_{T\xi} - \frac{\Delta_{T\xi}}{1-\xi_{S0}}(\xi_S - \xi_{S0}) \tag{1.23}$$

where if $T < T_0$,

$$\Delta_{T\xi} = \frac{1-\xi_{S0}-\xi_{T0}}{2}\cos\left[\frac{\pi}{M_s-M_f}(T-M_f)\right]+\frac{1-\xi_{S0}+\xi_{T0}}{2} \tag{1.24}$$

otherwise,

$$\Delta_{T\xi} = \xi_{T0} \tag{1.25}$$

Region 6: Transformation from austenite to twinned martensite

$$\xi_S = \xi_{S0} \tag{1.26}$$

$$\xi_T = \frac{1-\xi_0}{2}\cos\left[a_M(T-M_f)\right] + \frac{1+\xi_0}{2} - \xi_{S0} \tag{1.27}$$

Region 7: Transformation from twinned martensite to detwinned martensite

$$\xi_S = \frac{1-\xi_{S0}}{2}\cos\left[\frac{\pi}{\sigma_s^{cr}-\sigma_f^{cr}}(\sigma-\sigma_f^{cr})\right] + \frac{1+\xi_{S0}}{2} \tag{1.28}$$

$$\xi_T = \xi_{T0} - \frac{\xi_{T0}}{1-\xi_{S0}}(\xi_S-\xi_{S0}) \tag{1.29}$$

Region 8: Twinned and detwinned martensite are present, depending on the loading/temperature history; no transformation occurs

$$\xi_S = \xi_{S0} \tag{1.30}$$

$$\xi_T = \xi_{T0} \tag{1.31}$$

1.5.1 Examples

Using the model presented in this section, it is possible to compute the martensite fraction for various thermomechanical $\sigma-T$ paths. To allow a more comprehensive learning experience, a complete Wolfram Mathematica implementation of the model is available at http://smartsys.eng.utoledo.edu. The accompanying code allows for simulation of various $\sigma-T$ paths.

As the first example, we consider the superelastic effect, described by the path A–B–C–D–E–F–A in Figure 1.26. The point A is in the region 4, where $\xi_S = \xi_T = 0$ (Eqs. 1.20 and 1.21), so the alloy is completely in the austenite phase. By imposing a stress (tensile for a wire), the alloy crosses zone 4, zone 3 and zone 2, where no transformation occurs and therefore, stress-induced martensite and temperature-induced martensite are still both 0.

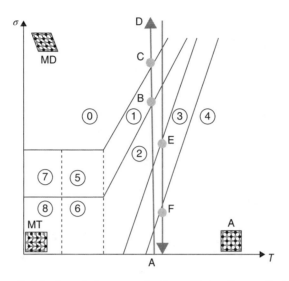

Figure 1.26 Example superelastic path on the $\sigma - T$ plane

(Eqs. 1.15–1.19). If the stress is further increased, it is possible to arrive at point B in zone 1, where the austenite phase starts to transform to detwinned martensite (Eqs. 1.13 and 1.14). The transformation occurs within the whole region up to point C, where $\xi_S = 1$, $\xi_T = 0$. These values of martensite fractions are considered as the initial values ($\xi_{S0} = 1$, $\xi_{T0} = 0$) in the transformation in the following zone (zone 0, Eqs. 1.11 and 1.12). After point C, the stress can be increased and no transformation occurs up to the martensite yield stress, where the plastic flow begins (not considered here). It is worth noting that the yield stress can occur only for stress induced martensite.

When the stress decreases (D–E), the zones 1 and 2 are crossed again, and no transformation occurs as $\xi_{S0} = 1$, $\xi_{T0} = 0$ (Eqs. 1.13–1.16). By further decreasing the stress, E–F, region 3 is crossed and the inverse transformation from martensite to austenite occurs (Eqs. 1.17–1.19) up to point F, where $\xi_S = 0$, $\xi_T = 0$. Finally, zone 4 is reached. Here, it is necessary to reupdate the initial values of martensite fractions to $\xi_{S0} = 0$, $\xi_{T0} = 0$ which remains constant in the zone, and the starting condition is restored (Eqs. 1.20 and 1.21).

As another example, we consider the SME, depicted by the path A–B–C–D–E–F–G–H–A in Figure 1.27. The point A is in the region 8 where, if previously no stress (for example tension for wire) is applied to the alloy, only twinned martensite is present (Eqs. 1.30 and 1.31, $\xi_S = 0$, $\xi_T = 1$). By applying tension, point B is reached where the martensite detwinning transformation begins.

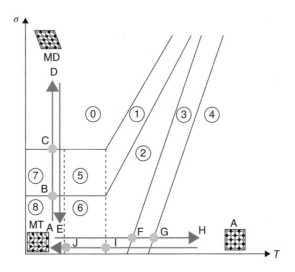

Figure 1.27 Shape memory effect path on the $\sigma - T$ plane

During the path B–C, the alloy is in the zone 7 where the stress-induced martensite increases, and the temperature-induced martensite decreases as the tension rises (Eqs. 1.28 and 1.29). Once point C is reached, only stress-induced martensite is present (Eqs. 1.11 and 1.12, $\xi_S = 1$, $\xi_T = 0$). If the stress is still increased, no transformation will occur up to the martensite yield stress, where the plastic flow begins. These martensite fraction values will be considered as the initial values for the transformation in zone 0 ($\xi_{S0} = 1$, $\xi_{T0} = 0$, path C–D) where they remain constant. By removing the stress, zone 7 and zone 8 are crossed where no transformation occurs (Eqs. 1.28–1.31). At the end of this first part of the path (A–B–C–D–E), martensite fractions are computed in zone 8, point E, with $\xi_{S0} = 1$, $\xi_{T0} = 0$.

Subsequently, the alloy is heated (A–F–G–H). Up to point F, zone 6 and zone 2 are crossed and no transformation occurs since $\xi_{S0} = 1$, $\xi_{T0} = 0$ (Eqs. 1.15 and 1.16 and Eqs. 1.26 and 1.27). Then zone 3 is crossed (F–G) and the inverse transformation occurs (from martensite to austenite; Eqs. 1.17–1.19) in zone 3. Once point G is reached, martensite has completely transformed to austenite and the martensite fraction values have to be considered as the initial values for the following zone (zone 4, $\xi_{S0} = 0$, $\xi_{T0} = 0$). Finally, the alloy is cooled following the path H–G–F–I–J–A. Up to point I, no transformation occurs because in zones 3 and 2 martensite fractions remain constant if $\xi_{S0} = 0$, $\xi_{T0} = 0$ (Eqs. 1.15–1.19). Then zone 6 is crossed where the austenite completely transforms to twinned martensite (Eqs. 1.26 and 1.27) and the initial condition is restored in zone 8 ($\xi_S = 0$, $\xi_T = 1$).

1.6 Summary

This chapter introduced the unique properties of SMAs such as the SME and superelasticity. The underlying microstructural phenomena corresponding to such behaviors were briefly discussed, and the phase diagram was introduced as a beneficial tool to schematically represent the associated phase transformations. As mentioned, although NiTi is the most common SMA, other shape memory materials are also available for specific applications. A well-known phenomenological modeling approach was presented for the design and analysis of the SMA actuators.

References

[1] Ölander A. The crystal structure of AuCd. Zeitschrift Fur Kristallographie 1932;**83**:145–8.

[2] Ölander A. An Electrochemical Investigation of Solid Cadmium–Gold Alloys. Journal of the American Chemical Society 1932;**54**:3819–33.

[3] Chang L, Read T. Plastic Deformation And Diffusionless Phase Changes In Metals—the Gold–Cadmium Beta-phase. Transactions of the American Institute of Mining and Metallurgical Engineers 1951;**191**:47–52.

[4] Anson T. Shape Memory Alloys—Medical Applications. Materials World 1999;**7**:745–7.

[5] Van Humbeeck J. Shape Memory Alloys: A Material and a Technology. Advanced Engineering Materials 2001;**3**:837–50.

[6] Buehler W, Gilfrich J, Wiley R. Effect of Low-temperature Phase Changes on the Mechanical Properties of Alloys Near Composition TiNi. Journal of Applied Physics 1963;**34**:1475–7.

[7] Wang F, Buehler W, Pickart S. Crystal Structure and a Unique Martensitic Transition of TiNi. Journal of Applied Physics 1965;**36**:3232–9.

[8] Brook G. Applications of Titanium–Nickel Shape Memory Alloys. Materials & Design 1983;**4**:835–40.

[9] Jee K, Han J, Jang W. A Method of Pipe Joining Using Shape Memory Alloys. Materials Science and Engineering A 2006;**438**:1110–2.

[10] Xu M, Song G. Adaptive Control of Vibration Wave Propagation in Cylindrical Shells Using SMA Wall Joint. Journal of Sound and Vibration 2004;**278**:307–26.

[11] Petrini L, Migliavacca F, Massarotti P, Schievano S, Dubini G, Auricchio F. Computational Studies of Shape Memory Alloy Behavior in Biomedical Applications. Journal of Biomechanical Engineering 2005;**127**:716–25.

[12] Taheri Andani M, Anderson W, Elahinia M. Design, Modeling and Experimental Evaluation of a Minimally Invasive Cage for Spinal Fusion Surgery Utilizing Superelastic Nitinol Hinges. Journal of Intelligent Material Systems and Structures 2015;**26**(6):631–638.

[13] Taheri Andani M, Elahinia M. Modeling and Simulation of SMA Medical Devices Undergoing Complex Thermo-Mechanical Loadings. Journal of Materials Engineering and Performance 2014;**23**:2574–83.

[14] Hartl D, Lagoudas D. Aerospace Applications of Shape Memory Alloys. Proceedings of the Institution of Mechanical Engineers, Part G: Journal of Aerospace Engineering 2007;**221**:535–52.

[15] DesRoches R, Smith B. Shape Memory Alloys in Seismic Resistant Design and Retrofit: a Critical Review of their Potential and Limitations. Journal of Earthquake Engineering 2004;**8**:415–29.

[16] Van Humbeeck J. Non-medical Applications of Shape Memory Alloys. Materials Science and Engineering A 1999;**273**:134–48.

[17] Morgan NB, Friend CM. A Review of Shape Memory Stability in NiTi Alloys. Le Journal de Physique IV 2001;**11**:Pr8-325–32.

[18] Eggeler G, Hornbogen E, Yawny A, Heckmann A, Wagner M. Structural and Functional Fatigue of NiTi Shape Memory Alloys. Materials Science and Engineering A 2004;**378**:24–33.

[19] Elahinia M, Hashemi M, Tabesh M, Bhaduri S. Manufacturing and Processing of NiTi Implants: a Review. Progress in Materials Science 2012;**57**:911–46.

[20] Duerig T, Pelton A, Stöckel D. An Overview of Nitinol Medical Applications. Materials Science and Engineering A 1999;**273**:149–60.

[21] Machado L, Savi M. Medical Applications of Shape Memory Alloys. Brazilian Journal of Medical and Biological Research 2003;**36**:683–91.

[22] Tarniţă D, Tarniţă DN, Bîzdoacă N, Mîndrilă I, Vasilescu M. Properties and Medical Applications of Shape Memory Alloys. Romanian Journal of Morphology and Embryology 2009;**50**:15–21.

[23] Taheri Andani M, Moghaddam N, Haberland C, Dean D, Miller M, Elahinia M. Metals for Bone Implants, Part 1: Powder Metallurgy and Implant Rendering. Acta Biomaterialia 2014;**10**(10):4058–70.

[24] Shimizu K, Tadaki T. Preface. Annu. Rev. Mater. Sci.;**18**(1).

[25] Hornbogen E, Warlimont H. Metallkunde: Aufbau und Eigenschaften von Metallen und Legierungen. Tokyo: Springer-Verlag GmbH; 2001.

[26] Saburi T, Tatsumi T, Nenno S. Effects of Heat Treatment on Mechanical Behavior of Ti–Ni Alloys. Le Journal de Physique Colloques 1982;**43**:c4-261–6.

[27] Otsuka K, Ren X. Physical Metallurgy of Ti–Ni-based Shape Memory Alloys. Progress in Materials Science 2005;**50**:511–678.

[28] Massalski T, Okamoto H, Subramanian P. Kacprzak L., editors. Binary Alloy Phase Diagrams, vols **1–3**. Materials Park, OH: ASM International; 1996.

[29] Hume-Rothery W, Poole D. Methods for Determining the Liquidus Points of Titanium-rich Alloys. Journal of the Institute of Metals 1954;**82**:490–2.

[30] Nishida M, Wayman C, Honma T. Phase Transformations in a $Ti_{50}Ni_{47.5}Fe_{2.5}$ Shape Memory Alloy. Metallography 1986;**19**:99–113.

[31] Zhang Z, Frenzel J, Somsen C, Pesicka J, Neuking K, Eggeler G. Orientation Relationship between TiC Carbides and B2 Phase in As-cast and Heat-treated NiTi Shape Memory Alloys. Materials Science and Engineering A 2006;**438**:879–82.

[32] Bastin G, Rieck G. Diffusion in the Titanium–nickel System: I. Occurrence and Growth of the Various Intermetallic Compounds. Metallurgical Transactions 1974;**5**:1817–26.

[33] Khalil-Allafi J, Dlouhy A, Eggeler G. N_4Ti_3-Precipitation During Aging of NiTi Shape Memory Alloys and its Influence on Martensitic Phase Transformations. Acta Materialia 2002;**50**:4255–74.

[34] Khalil-Allafi J, Eggeler G, Dlouhy A, Schmahl W, Somsen C. On the Influence of Heterogeneous Precipitation on Martensitic Transformations in a Ni-rich NiTi Shape Memory Alloy. Materials Science and Engineering A 2004;**378**:148–51.

[35] Ortega A, Tyber J, Frick C, Gall K, Maier H. Cast NiTi Shape-Memory Alloys. Advanced Engineering Materials 2005;**7**:492–507.

[36] Wagner M, Eggeler G. New Aspects of Bending Rotation Fatigue in Ultra-fine-grained Pseudo-elastic NiTi Wires: Dedicated to Professor Eckard Macherauch on the Occasion of the 80th Anniversary of his Birth. Zeitschrift für Metallkunde 2006;**97**:1687–96.

[37] Miyazaki S, Imai T, Igo Y, Otsuka K. Effect of Cyclic Deformation on the Pseudoelasticity Characteristics of Ti–Ni Alloys. Metallurgical Transactions A 1986;**17**:115–20.

[38] Tang W, Sundman B, Sandström R, Qiu C. New Modelling of the B2 Phase and its Associated Martensitic Transformation in the Ti–Ni System. Acta Materialia 1999;**47**:3457–68.

[39] Frenzel J, George E, Dlouhy A, Somsen C, Wagner M, Eggeler G. Influence of Ni on Martensitic Phase Transformations in NiTi Shape Memory Alloys. Acta Materialia 2010;**58**:3444–58.

[40] Tong H, Wayman C. Characteristic Temperatures and Other Properties of Thermoelastic Martensites. Acta Metallurgica 1974;**22**:887–96.

[41] Saburi T. Ti–Ni Shape Memory Alloys. In: Otsuka K, Wayman CM, editors. Shape Memory Materials. Cambridge, UK: Cambridge University Press; 1999, pp. 49–96.

[42] Zhang Z, Frenzel J, Somsen C, Pesicka J, Eeggeler G. On the Formation of TiC Crystals During Processing of NiTi Shape Memory Alloys. In: Karas G, editor. Trends in Crystal Growth Research. New York: Nova Science Publishers; 2005, pp. 71–99.

[43] Frenzel J, Zhang Z, Somsen C, Neuking K, Eggeler G. Influence of Carbon on Martensitic Phase Transformations in NiTi Shape Memory Alloys. Acta Materialia 2007;**55**:1331–41.

[44] Frenzel J, Neuking K, Eggeler G, Haberland C. On the Role of Carbon During Processing of NiTi Shape Memory Alloys. In: Miyazaki S, editor. Proceedings

of the International Conference on Shape Memory and Superelastic Technologies. Tsukuba: ASM International; 2007, pp. 131–8.

[45] Nevitt M. Stabilization of Certain Ti$_2$Ni-type Phases by Oxygen Transactions of the Metallurgical Society of AIME 1960;**218**:327–31.

[46] Olier P, Barcelo F, Bechade J, Brachet J, Lefevre E, Guenin G. Effects of Impurities Content (oxygen, carbon, nitrogen) on Microstructure and Phase Transformation Temperatures of Near Equiatomic TiNi Shape Memory Alloys. Le Journal de Physique IV 1997;**7**:C5-143–8.

[47] Bekker A, Brinson C. Phase Diagram Based Description of the Hysteresis Behavior of Shape Memory Alloys. Acta Materialia 1998;**46**:3649–65.

[48] Šittner P, Takakura M, Tokuda M. Shape Memory Effects Under Combined Forces. Materials Science and Engineering A 1997;**234**:216–9.

[49] Miyazaki S, Igo Y, Otsuka K. Effect of Thermal Cycling on the Transformation Temperatures of Ti–Ni Alloys. Acta Metallurgica 1986;**34**:2045–51.

[50] Johnson W, Domingue J, Reichman S. P/M Processing and Characterization of Controlled Transformation Temperature NiTi. Le Journal de Physique Colloques 1982;**43**:c1-285–90.

[51] Madangopal K, Banerjee S, Lele S. Thermal Arrest Memory Effect. Acta Metallurgica et Materialia 1994;**42**:1875–85.

[52] Airoldi G, Corsi A, Riva G. The Hysteresis Cycle Modification in Thermoelastic Martensitic Transformation of Shape Memory Alloys. Scripta Materialia 1997;**36**:1273–8.

[53] Airoldi G, Corsi A, Riva G. Step-wise Martensite to Austenite Reversible Transformation Stimulated by Temperature or Stress: a Comparison in NiTi Alloys. Materials Science and Engineering A 1998;**241**:233–40.

[54] Wang Z, Zu X, Fu Y. Review on the Temperature Memory Effect in Shape Memory Alloys. International Journal of Smart and Nano Materials 2011;**2**:101–19.

[55] Zheng Y, Cui L, Schrooten J. Temperature Memory Effect of a Nickel–Titanium Shape Memory Alloy. Applied Physics Letters 2004;**84**:31–3.

[56] Buravalla V, Khandelwal A. Evolution Kinetics in Shape Memory Alloys Under Arbitrary Loading: Experiments and Modeling. Mechanics of Materials 2011;**43**:807–23.

[57] Birnbaum AJ, Yao Y. The Effects of Laser Forming on NiTi Superelastic Shape Memory Alloys. Journal of Manufacturing Science and Engineering 2010;**132**:041002.

[58] Khan M, Pequegnat A, Zhou YN. Multiple Memory Shape Memory Alloys. Advanced Engineering Materials 2013;**15**:386–93.

[59] Tang C, Huang WM, Wang C, Purnawali H. The Triple-Shape Memory Effect in NiTi Shape Memory Alloys. Smart Materials and Structures 2012;**21**:085022.

[60] Barth J, Krevet B, Kohl M. A Bistable Shape Memory Microswitch with High Energy Density. Smart Materials and Structures 2010;**19**:094004.

[61] Makino E, Mitsuya T, Shibata T. Fabrication of TiNi Shape Memory Micropump. Sensors and Actuators A: Physical 2001;**88**:256–62.

[62] Paik JK, Hawkes E, Wood R. A Novel Low-profile Shape Memory Alloy Torsional Actuator. Smart Materials and Structures 2010;**19**:125014.

[63] Kohl M, Skrobanek K. Linear Microactuators Based on the Shape Memory Effect. Sensors and Actuators A: Physical 1998;**70**:104–11.

[64] Elzey D, Sofla A, Wadley H. A Bio-inspired High-authority Actuator for Shape Morphing Structures. Proceedings of the SPIE 5053, Smart Structures and Materials 2003: Active Materials: Behavior and Mechanics, 92 (August 12, 2003); doi:10.1117/12.484745.

[65] Sofla A, Elzey D, Wadley H. Two-way Antagonistic Shape Actuation Based on the one-way Shape Memory Effect. Journal of Intelligent Material Systems and Structures 2008;**19**(9):1017–1027.

[66] Yamauchi K, Ohkata I, Tsuchiya K, Miyazaki S. Shape Memory and Superelastic Alloys: Applications and Technologies. Cambridge, UK: Elsevier; 2011.

[67] Hartl D, Mooney J, Lagoudas D, Calkins F, Mabe J. Use of a $Ni_{60}Ti$ Shape Memory Alloy for Active jet Engine Chevron Application: II. Experimentally validated numerical analysis. Smart Materials and Structures 2010;**19**:015021.

[68] Mabe J, Ruggeri R, Rosenzweig E. NiTinol Performance Characterization and Rotary Actuator Design. Proceedings of the SPIE 5388, Smart Structures and Materials 2004: Industrial and Commercial Applications of Smart Structures Technologies, 95 (July 29, 2004); doi:10.1117/12.539008.

[69] Calkins F, Mabe J. Shape Memory Alloy based Morphing Aerostructures. Journal of Mechanical Design 2010;**132**:111012.

[70] Mabe J, Brown J, Calkins F. Flight Test of a Shape Memory Alloy Actuated Adaptive Trailing Edge Flap, Part 1. The International Conference on Shape Memory and Superelastic Technologies (SMST) (May 12–16, 2014), Pacific Grove, CA, USA. ASM; 2014.

[71] Calkins F, Mabe J, Brown J. Flight Test of a Shape Memory Alloy Actuated Adaptive Trailing Edge Flap, Part 2. The International Conference on Shape Memory and Superelastic Technologies (SMST) (May 12–16, 2014), Pacific Grove, CA, USA. ASM; 2014.

[72] Langbein S, Czechowicz A. Problems and Solutions for Shape Memory Actuators in Automotive Applications. ASME 2012 Conference on Smart Materials, Adaptive Structures and Intelligent Systems (September 19–21, 2012), Stone Mountain, GA, USA. American Society of Mechanical Engineers; 2012, pp. 433–9.

[73] Czechowicz A, Lygin K, Langbein S. On the Potentials of Shape Memory Alloy Valves. Journal of Materials Engineering and Performance 2014;**23**:2687–95.

[74] Mammano GS, Dragoni E. Modelling of Wire-on-drum Shape Memory Actuators for Linear and Rotary Motion. Journal of Intelligent Material Systems and Structures 2011;**22**:1129–40.

[75] Dragoni E, Mammano GS. Modelling and Validation of a Rotary Motor Combining Shape Memory Wires and Overrunning Clutches. ASME Conference on Smart Materials, Adaptive Structures and Intelligent Systems (September 8–10, 2014), Newport, RI, USA. ASME; 2014.

[76] Kaneko K, Enomoto K. Development of Reciprocating Heat Engine Using Shape Memory Alloy. Journal of Environment and Engineering 2011;**6**:131–9.

[77] Elahinia M, Ashrafiuon H. Nonlinear Control of a Shape Memory Alloy Actuated Manipulator. Journal of Vibration and Acoustics 2002;**124**:566–75.

[78] Williams EA, Shaw G, Elahinia M. Control of An Automotive Shape Memory Alloy Mirror Actuator. Mechatronics 2010;**20**:527–34.

[79] Wang Z, Hang G, Wang Y, Li J, Du W. Embedded SMA Wire Actuated Biomimetic Fin: a Module for Biomimetic Underwater Propulsion. Smart Materials and Structures 2008;**17**:025039.

[80] Daghia F, Inman DJ, Ubertini F, Viola E. Shape Memory Alloy Hybrid Composite Plates for Shape and Stiffness Control. Journal of Intelligent Material Systems and Structures 2007;**19**:609–19.

[81] Furst S, Bunget G, Seelecke S. Design and Fabrication of a Bat-inspired Flapping-flight Platform Using Shape Memory Alloy Muscles and Joints. Smart Materials and Structures 2013;**22**:014011.

[82] Moghaddam M, Hadi A, Tohidi A, Elahinia M. Design, Modeling, and Prototyping of a Simple Semi-modular Biped Actuated by Shape Memory Alloys. Journal of Intelligent Material Systems and Structures 2011;**22**:1489–99.

[83] Malukhin K, Ehmann K. Material Characterization of NiTi Based Memory Alloys Fabricated by the Laser Direct Metal Deposition Process. Journal of Manufacturing Science and Engineering 2006;**128**:691–6.

[84] Elahinia M, Andani MT, Haberland C. Shape Memory and Superelastic Alloys. In: Bar-Cohen Y, editor. High Temperature Materials and Mechanisms. Boca Raton, FL: CRC Press; 2014, pp. 355–80.

[85] Elahinia M. Effect of System Dynamics on Shape Memory Alloy Behavior and Control. Blacksburg, VA: Virginia Tech; 2004.

[86] Elahinia M, Seigler T, Leo D, Ahmadian M. Nonlinear Stress-based Control of a Rotary SMA-actuated Manipulator. Journal of Intelligent Material Systems and Structures 2004;**15**:495–508.

[87] Elahinia M, Ahmadian M, Ashrafiuon H. Design of a Kalman Filter for Rotary Shape Memory Alloy Actuators. Smart Materials and Structures 2004;**13**:691.

[88] Elahinia M, Koo J, Ahmadian M, Woolsey C. Backstepping Control of a Shape Memory Alloy Actuated Robotic Arm. Journal of Vibration and Control 2005; **11**:407–29.

[89] Elahinia M, Ashrafiuon H, Ahmadian M, Tan H. A temperature-based controller for a shape memory alloy actuator. Journal of Vibration and Acoustics 2005; **127**:285–91.

[90] Elahinia M, Ahmadian M. An Enhanced SMA Phenomenological Model: I. The Shortcomings of the Existing Models. Smart Materials and Structures 2005;**14**:1297.

[91] Elahinia M, Ahmadian M. An Enhanced SMA Phenomenological Model: II. The Experimental Study. Smart Materials and Structures 2005;**14**:1309.

[92] Ashrafiuon H, Eshraghi M, Elahinia M. Position Control of a Three-link Shape Memory Alloy Actuated Robot. Journal of Intelligent Material Systems and Structures 2006;**17**:381–92.

[93] Elahinia M, Ahmadian M. Application of the Extended Kalman Filter to Control of a Shape Memory Alloy Arm. Smart Materials and Structures 2006;**15**:1370.

[94] Tanaka K. A Thermomechanical Sketch of Shape Memory Effect: One-dimensional Tensile Behavior. Res Mechanica 1986;**18**:251–63.

[95] Brinson L. One-dimensional Constitutive Behavior of Shape Memory Alloys: Thermomechanical Derivation with Non-constant Material Functions and Redefined Martensite Internal Variable. Journal of Intelligent Material Systems and Structures 1993;**4**:229–42.

[96] Liang C, Rogers C. One-dimensional Thermomechanical Constitutive Relations for Shape Memory Materials. Journal of Intelligent Material Systems and Structures 1990;**1**:207–34.

[97] Williams E, Elahinia MH. An Automotive SMA Mirror Actuator: Modeling, Design, and Experimental Evaluation. Journal of Intelligent Material Systems and Structures 2008;**19**:1425–34.

[98] Zaidi S, Lamarque F, Prelle C, Carton O, Zeinert A. Contactless and Selective Energy Transfer to a Bistable Micro-actuator Using Laser Heated Shape Memory Alloy. Smart Materials and Structures 2012;**21**:115027.

[99] Chung J-H, Heo J-S, Lee J-J. Implementation Strategy for the Dual Transformation Region in the Brinson SMA Constitutive Model. Smart Materials and Structures 2007;**16**:N1–5.

2

Mathematical Modeling and Simulation

Reza Mirzaeifar and Mohammad H. Elahinia

The objective of this chapter is studying mathematical modeling and simulation of SMAs, particularly nickel–titanium (NiTi) alloys at various length scales. The following sections mainly cover two different scales and introduce appropriate strategies for studying the thermomechanical response of the material in each scale.

In macroscale, and for most of the engineering applications, phenomenological constitutive models are implemented into numerical frameworks for studying various geometries subjected to different loading conditions. Although the existing numerical methods are efficient for studying a large number of applied SMA devices, there are also several cases in which these numerical frameworks suffer from the high computational cost and convergence difficulties (i.e., when the geometric nonlinearities are not negligible). Also, results obtained using these numerical methods are highly sensitive to a large number of secondary parameters, for example, the mesh size, tolerance criteria, and number of loading steps. The numerical simulations may give erroneous results due to an improper choice for any of these parameters. In this chapter, we study some closed-form solution for analyzing the

Shape Memory Alloy Actuators: Design, Fabrication, and Experimental Evaluation, First Edition. Mohammad H. Elahinia.
© 2016 John Wiley & Sons, Ltd. Published 2016 by John Wiley & Sons, Ltd.

superelastic response of shape memory alloys in various loading conditions. The closed-form (or semianalytic) solutions are obtained by reducing the three-dimensional constitutive models to appropriate one- or two-dimensional constitutive equations. Closed-form expressions are obtained for the martensitic volume fraction and stress distributions. In addition to the J_2-based models that predict a symmetric response in tension and compression, the closed-form solutions can also be developed based on more accurate J_2–I_1 transformation functions which are capable of considering the tension–compression asymmetry (a well-known response for most SMAs [1, 2]). In the J_2-based models, the transformation function is expressed as a function of the second deviatoric stress invariant. In the J_2–I_1 models for describing the transformation function, in addition to the second deviatoric stress invariant, an extra term containing the first stress invariant is also considered. Another advantage of using closed-form solutions is the potential of these models to consider the coupled thermomechanical behavior of various large polycrystalline SMA devices subjected to different loadings, including uniaxial loads, torsion, and bending, as it is shown in this chapter.

SMAs have recently attracted considerable interest for applications as actuators in microelectromechanical systems (MEMS) [3–5] due to their relatively high work output per unit volume [6]. The mechanical properties of some different shape memory alloys have been studied experimentally at microscale by testing micropillars produced by focused ion beam (FIB) micromachining [7]. Recently, the pseudoelasticity, crystal orientation effect, and size dependency have been extensively studied experimentally for nickel–titanium and Cu–Al–Ni micropillars [8–12]. In order to have a precise description of the SMA microscaled devices, it is ideal to extract the material response in several loading conditions such as tension, compression, and bending experimentally. However, performing experimental measurements are complicated and costly in most of loading conditions. For example, among the reported works on studying the shape memory micropillar response, the majority of experiments are performed for compressive loading [8–11]; there are very few experimental works on bending [12]. While performing bending tests on micropillars, one faces some technical difficulties [12]. Tensile tests in the nano- and microscales are considerably more difficult because a special geometric shape should be created at the pillar head for attaching the tensile tool to the pillar [13, 14]. Motivated by these experimental difficulties, in the following sections, we introduce some micromechanical-based models for studying SMAs at microscale.

While NiTi has been considered as a potential smart material in various nanodevices recently, fundamental understanding of phase transformations in NiTi nanostructures remains largely unexplored. Although the martensitic phase transformation in bulk NiTi has been intensively studied, it still remains unclear how the surface affects phase transformation in nanostructured NiTi in NEMS applications, despite some recent efforts [15–20]. The structure transformation and thermomechanical behavior of nanoscale materials can be remarkably different from their bulk counterparts [21–24]. For studying some aspects of the thermomechanical properties of SMAs at nanoscale, molecular dynamics simulations can be used to explore the martensitic phase transformation in NiTi alloys at the atomistic level. The martensite reorientation, austenite to martensite phase transformation, and twinning mechanisms in NiTi nanostructures can be analyzed, and the effect of various parameters including the temperature and size on the phase transformation at the atomistic level can be studied [25].

Besides atomistic modeling, the time-independent and time-dependent Ginzburg–Landau phase field models are other alternatives for studying the martensitic phase transformation at the nanoscale. In physics, the Ginzburg–Landau theory, named after Vitaly Lazarevich Ginzburg and Lev Landau, is a mathematical theory originally used to model superconductivity. It has been shown that this theory can also be used for modeling the solid–solid phase transformation. The martensitic transformation is a spontaneous crystal lattice rearrangement driven by a reduction of the partial Gibbs free energy of the parent phase in the stress-free state. A comprehensive model of the martensitic transformation should explicitly take into account the transformation-induced strain and allow the system to follow any possible path without any preimposed constraints on the morphology of the martensite structure. The continuum field (phase field) methods based on the phenomenological Ginzburg–Landau equations seem to provide one of the most efficient tools to solve this problem. These methods provide an optimum combination of the generality of the approach and computational efficiency [26]. One of the early works that introduced the Ginzburg–Landau theory of phase transition for studying shape memory alloys is reported by Falk [27]. In this work, based on the Landaus theory, a phenomenological model free energy function is presented which accounts for the shape memory effect. The main idea in the early works on phase field models has been used in numerous researches in recent years for introducing various improvements of the Landau and Ginzburg–Landau theories for studying shape memory alloys [26, 28–33]. Most of the recent works on improving the phase field models for studying SMAs can be classified into the following major categories: (i) developing the available one-dimensional formulations to more realistic three-dimensional models, (ii) developing models capable of considering multivariant martensitic transformation, and (iii) considering the time-dependent models for studying kinetics of transformation and propagation of phase transformation boundaries.

2.1 Phenomenological Macroscale Modeling

The response of an SMA single crystal is distinctly different from polycrystalline SMAs. Besides the micromechanical approaches for developing SMA constitutive relations for modeling the behavior of single crystals, it is necessary to develop appropriate phenomenological methods for studying the macroscopic response of SMA polycrystalline devices. A polycrystalline SMA consists of many grains with different crystallographic orientations. The phase transformation strongly depends on the crystallographic orientation, and modeling the macroscopic response of SMAs by considering different phase transformation conditions in each grain is extremely difficult for a cluster of thousands of grains. Considering the complexity of microstructure-based modeling in large polycrystalline SMA devices, one is forced to use macroscopic phenomenological constitutive equations for modeling the martensitic transformation. These models are based on continuum thermo-mechanics and construct a macroscopic free energy potential (Helmholtz or Gibbs free energy) depending on the state and internal variables used to describe the measure of phase transformation. Consequently, evolution equations are postulated for the internal variables, and the second law of thermodynamics is used in order to find thermodynamic constraints on the material constitutive equations. In recent years, different constitutive models have been introduced by different choices of thermodynamic potentials, internal state variables, and their evolution equations (e.g., see [34–36] for a comprehensive list of one-dimensional and three-dimensional phenomenological SMA constitutive equations with different choices of thermodynamic potentials and internal state variables). Besides different choices of potential energy and internal state variables, by considering the experimental results for the response of SMAs, various choices have been made for the hardening function. Among the most widely accepted models, we can mention the cosine model [37], the exponential model [38], and the polynomial model [39]. These models are also unified using a thermodynamic framework [40]. In this chapter, for the macroscopic continuum-based modeling, we use the phenomenological constitutive equations using the Gibbs free energy as the thermodynamic potential, the martensitic volume fraction and transformation strains as the internal state variables, and the hardening function in polynomial form.

Among the solutions presented for studying polycrystalline SMAs subjected to various loadings, a large number are purely numerical [41–43]. As mentioned earlier, the existing numerical methods of modeling SMAs have several drawbacks including the high computational cost and convergence difficulties. On the other hand, the analytic and semianalytic solutions in the literature are mostly based on oversimplified constitutive relations or use of unrealistic simplifying assumptions. In this chapter, we introduce a set of closed-form solutions for analyzing the superelastic shape memory alloys subjected to various loadings. In each case, the general three-dimensional phenomenological constitutive model is reduced to an appropriate constitutive equation for the corresponding loading condition. Closed-form expressions are obtained for the martensitic volume fraction and stress distributions, and the superelastic response of the SMA device is studied analytically. Comparing the results of the closed-form solutions with several experimental and numerical results shows the accuracy of the proposed solution in each case.

Another advantage of developing the closed-form solutions is the ability of studying two complicated responses in SMAs: the thermomechanical coupling and the tension–compression asymmetry. It is worth noting that both of these effects can be studied with the numerical simulations as well. However, the computational cost and convergence difficulties increase remarkably by adding each of these effects to the numerical models. As will be shown in this chapter, the introduced closed-form solutions are capable of considering both these effects in various conditions. The closed-form solutions for the pure torsion of circular SMA bars, pure torsion of curved SMA bars, SMA helical springs, and bending of SMA beams are shown as representative examples. The thermomechanical coupling effect is also studied.

2.1.1 Three-Dimensional Macroscopic Phenomenological Constitutive Equations

In order to develop closed-form solutions for various SMA devices, we use the three-dimensional phenomenological macroscopic constitutive model for polycrystalline SMAs [39]. In this constitutive model, by considering the transformation strain ϵ^t (the portion of strain that is recovered due to reverse phase transformation from detwinned martensite to austenite) and the martensitic volume fraction ξ (an indicator of the extent of the phase transformation from austenite to martensite) as the internal state variables, the following expression is obtained for the Gibbs free energy potential [44]:

$$
G(\boldsymbol{\sigma}, T, \epsilon^t, \xi) = -\frac{1}{2\rho}\boldsymbol{\sigma}:\mathbb{S}:\boldsymbol{\sigma} - \frac{1}{\rho}\boldsymbol{\sigma}:\left[\alpha(T-T_0)+\epsilon^t\right] + c\left[(T-T_0)-T\ln\left(\frac{T}{T_0}\right)\right]
$$
$$
-s_0 T + u_0 + \frac{1}{\rho}f(\xi) \tag{2.1}
$$

where $\mathbb{S}$, α, c, ρ, s_0, and u_0 are the effective compliance tensor, effective thermal expansion coefficient tensor, effective specific heat, mass density, effective specific entropy, and effective specific internal energy at the reference state, respectively. The symbols $\boldsymbol{\sigma}$, T, T_0, ϵ^t, and ξ denote the Cauchy stress tensor, temperature, reference temperature, transformation strain, and martensitic volume fraction, respectively. Any effective material property $\mathbf{P}$ is assumed to vary with the martensitic volume fraction as $\mathbf{P} = \mathbf{P}^A + \xi\Delta\mathbf{P}$, where the superscript A denotes the austenite phase and the symbol $\Delta(.)$ denotes the difference of a quality (.) between the martensitic and austenitic phases, that is, $\Delta(.) = (.)^M - (.)^A$ with M denoting the martensite phase.

In (2.1), $f(\xi)$ is a hardening function that models the transformation strain hardening in the SMA material. Several different choices have been introduced for this function. Here, we use a polynomial hardening model as [36]

$$
f(\xi) = \begin{cases} \frac{1}{2}\rho b^M \xi^2 + (\mu_1+\mu_2)\xi, & \dot{\xi} > 0, \\ \frac{1}{2}\rho b^A \xi^2 + (\mu_1-\mu_2)\xi, & \dot{\xi} < 0 \end{cases} \tag{2.2}
$$

where ρb^A, ρb^M, μ_1, and μ_2 are material constants for transformation strain hardening. The first condition in (2.2) represents the forward phase transformation $(A \rightarrow M)$, and the second condition represents the reverse phase transformation $(M \rightarrow A)$. The constitutive relation of a shape memory material can be obtained by using the total Gibbs free energy as

$$
\epsilon = -\rho\frac{\partial G}{\partial \boldsymbol{\sigma}} = \mathbb{S}:\boldsymbol{\sigma} + \alpha(T-T_0) + \epsilon^t \tag{2.3}
$$

where ϵ is the strain tensor. By introducing a generalized thermodynamic force $\mathcal{P}$ as

$$
\mathcal{P} = -\rho\frac{\partial G}{\partial \xi} = \frac{1}{2}\boldsymbol{\sigma}:\Delta\mathbb{S}:\boldsymbol{\sigma} + \Delta\alpha:\boldsymbol{\sigma}(T-T_0) + \rho\Delta c\left[(T-T_0)-T\ln\left(\frac{T}{T_0}\right)\right]
$$
$$
+ \rho\Delta s_0 T - \frac{\partial f}{\partial \xi} - \rho\Delta u_0 \tag{2.4}
$$

the second law of thermodynamics in the form of nonnegativeness of the rate of entropy production density can be expressed as $\boldsymbol{\sigma} : \dot{\boldsymbol{\epsilon}}^t + \mathcal{P}\dot{\xi} = \pi\dot{\xi} \geq 0$ [45]. We assume the existence of a thermoelastic region (transformation surface) bounded by a smooth hypersurface, which can be described by a transformation function Φ as $\Phi(\boldsymbol{\sigma}, \mathcal{P}) = 0$. We choose the following general form for the transformation function [44]:

$$\Phi(\boldsymbol{\sigma}, \mathcal{P}) = \left[\tilde{\Phi}(\boldsymbol{\sigma}) + \mathcal{P}\right]^2 - Y^2 = \left[\tilde{\Phi}(\boldsymbol{\sigma}) + \mathcal{P} + Y\right]\left[\tilde{\Phi}(\boldsymbol{\sigma}) + \mathcal{P} - Y\right] \qquad (2.5)$$

where $\tilde{\Phi}(\boldsymbol{\sigma})$ is the stress-related transformation function that will be defined in the sequel and Y is a measure of internal dissipation due to microstructural changes during phase transformation. The transformation surface that controls the onset of direct (austenite to martensite) and reverse (martensite to austenite) phase transformation is defined as

$$\tilde{\Phi}(\boldsymbol{\sigma}) + \mathcal{P} = \begin{cases} Y, \dot{\xi} > 0, \\ -Y, \dot{\xi} < 0 \end{cases} \qquad (2.6)$$

Considering the fact that any change in the state of the system is only possible by a change in the internal state variable ξ [46], the evolution of the transformation strain tensor is related to the evolution of the martensitic volume fraction as $\dot{\boldsymbol{\epsilon}}^t = \left(\partial\tilde{\Phi}(\boldsymbol{\sigma})/\partial\boldsymbol{\sigma}\right)\dot{\xi} = \boldsymbol{\Gamma}\dot{\xi}$, where $\boldsymbol{\Gamma}$ represents a transformation tensor associated with the chosen transformation function, related to the deviatoric stress tensor as

$$\boldsymbol{\Gamma} = \begin{cases} \dfrac{3H}{2\bar{\sigma}}\boldsymbol{\sigma}', \dot{\xi} > 0, \\[2mm] \dfrac{H}{\bar{\epsilon}^{tr}}\boldsymbol{\epsilon}^{tr}, \dot{\xi} < 0 \end{cases} \qquad (2.7)$$

In (2.7), H is the maximum uniaxial transformation strain and ϵ^{tr} represents the transformation strain at the reverse phase transformation. The terms $\boldsymbol{\sigma}'$, $\bar{\sigma}$, and $\bar{\epsilon}^{tr}$ are the deviatoric stress tensor, the second deviatoric stress invariant, and the second deviatoric transformation strain invariant, respectively.

The closed-form solution presented in the following sections are focused on studying the pseudoelastic response of SMAs. However, since all the equations are considered in a general form by keeping the temperature as an independent parameter, the set of introduced solution can be easily developed for studying the shape memory effect as well.

2.1.2 Closed-Form Solutions for Pure Torsion of Shape Memory Alloy Circular Bars

As the first case, we consider the pure torsion of SMA circular bars. Developing closed-form solutions for this loading condition enables us to model SMA helical springs, as presented in the next section, which has several applications as smart actuators. The closed-form solution for pure torsion of shape memory alloy circular bars ignoring the effect of phase transformation latent heat and assuming the isothermal condition can be obtained by reducing the three-dimensional constitutive model to an appropriate one-dimensional case, using the fact that stress, strain, and transformation strain tensors have only one non-zero term. Simplifying the second deviatoric stress and transformation strain invariants and replacing them into (2.7) simplify the transformation tensor to

$$\mathbf{\Gamma}^{+} = \frac{\sqrt{3}}{2} H \operatorname{sgn}(\tau_{\theta z}) \begin{bmatrix} 0 & 0 & 0 \\ 0 & 0 & 1 \\ 0 & 1 & 0 \end{bmatrix}, \mathbf{\Gamma}^{-} = \frac{\sqrt{3}}{2} H \operatorname{sgn}(\epsilon_{\theta z}^{tr}) \begin{bmatrix} 0 & 0 & 0 \\ 0 & 0 & 1 \\ 0 & 1 & 0 \end{bmatrix} \quad (2.8)$$

where $\tau_{\theta z}$ is the shear stress, $\epsilon_{\theta z}^{tr}$ is the transformation shear strains, sgn(.) is the sign function, and the superscripts $+$ and $-$ for $\mathbf{\Gamma}$ represent the forward and inverse phase transformations, respectively. Substituting (2.8) into (2.4) and (2.6) gives the following explicit expressions for martensitic volume fractions in direct and inverse phase transformations (see Ref. [47] for more details):

$$\xi^{+} = \frac{1}{\rho b^{M}} \left\{ \sqrt{3} H |\tau_{\theta z}| + 2\tau_{\theta z}^{2} \Delta S_{44} + f^{+}(T) \right\}, \quad (2.9)$$

$$\xi^{-} = \frac{1}{\rho b^{A}} \left\{ \sqrt{3} H \tau_{\theta z} \operatorname{sgn}(\epsilon_{\theta z}^{tr}) + 2\tau_{\theta z}^{2} \Delta S_{44} + f^{-}(T) \right\} \quad (2.10)$$

where

$$f^{+}(T) = \rho \Delta c \left[(T - T_0) - T \ln\left(\frac{T}{T_0}\right) \right] + \rho \Delta s_0 (T - M_s), \quad (2.11)$$

$$f^{-}(T) = \rho \Delta c \left[(T - T_0) - T \ln\left(\frac{T}{T_0}\right) \right] + \rho \Delta s_0 (T - A_f) \quad (2.12)$$

The parameters M_s and A_f are the martensitic start and austenite finish temperatures, respectively. By substituting the explicit expression of the martensitic volume fraction into the evolution equation and after integrating from zero to an arbitrary time, the transformation shear strain can be calculated. The constitutive relation (2.3) is now reduced to read

$$\epsilon_{\theta z} = \frac{1+\nu}{E_A + \xi^\pm (E_M - E_A)} \tau_{\theta z}$$
$$+ \frac{1}{\rho b^\pm} \left\{ \frac{3}{2} H^2 \tau_{\theta z} + \sqrt{3} H \tau_{\theta z}^2 \aleph^\pm \Delta S_{44} + \frac{\sqrt{3}}{2} H \aleph^\pm f^\pm (T) \right\} \tag{2.13}$$

for the pure torsion, where the + and − symbols are used for the direct and reverse phase transformations, respectively, and the other parameters are $\aleph^+ = \mathrm{sgn}(\tau_{\theta z})$, $\aleph^- = \mathrm{sgn}(\epsilon_{\theta z}^{\mathrm{tr}})$, $\rho b^+ = \rho b^M$, and $\rho b^- = \rho b^A$. For a bar with a circular cross section, the shear strain in (2.13) can be related to twist angle per unit length as $\epsilon_{\theta z} = \frac{1}{2} r \theta$, where r is the distance from the axis of the bar. Substituting (2.9) into (2.13) and considering the special case in which both the shear stress and the shear transformation strains are positive, (2.13) can be rewritten as

$$\tau_{\theta z}^4 + F_1 \tau_{\theta z}^3 + (F_2 + F_2^* r\theta) \tau_{\theta z}^2 + (F_3 + F_3^* r\theta) \tau_{\theta z} + (F_4 + F_4^* r\theta) = 0 \tag{2.14}$$

where

$$F_1 = \frac{\sqrt{3}H}{\Delta S_{44}},$$

$$F_2 = \frac{1}{4} \frac{3\Delta E H^2 + 4\Delta E \Delta S_{44} f^\pm(T) + 2\rho b^\pm E_A \Delta S_{44}}{\Delta E \Delta S_{44}^2},$$

$$F_3 = \frac{\sqrt{3}}{12} \frac{6\Delta E H^2 f^\pm(T) + 3\rho b^\pm E_A H^2 + 2(1+\nu)(\rho b^\pm)^2}{H \Delta E \Delta S_{44}^2},$$

$$F_4 = \frac{1}{4} \frac{f^\pm(T)(\Delta E f^\pm(T) + \rho b^\pm E_A)}{\Delta E \Delta S_{44}^2}, \tag{2.15}$$

$$F_2^* = -\frac{1}{6} \frac{\rho b^\pm \sqrt{3}}{\Delta S_{44} H},$$

$$F_3^* = -\frac{1}{4} \frac{\rho b^\pm}{\Delta S_{44}^2},$$

$$F_4^* = -\frac{\sqrt{3}}{12} \frac{\rho b^\pm (\Delta E f^\pm(T) + \rho b^\pm E_A)}{H \Delta E \Delta S_{44}^2}$$

in which $\Delta E = (E_M - E_A)$. The relation (2.14) is a quartic equation that can be solved analytically using Ferrari's method[1] for finding the shear stress $\tau_{\theta z}$ as a function of twist angle in an arbitrary radius as

$$\tau_{\theta z} = \wp^{\pm}(r,\theta) = -\frac{1}{4}F_1 + \frac{1}{2}\mathcal{W} - \frac{1}{2}\sqrt{-3\alpha - 2\mathcal{Y} - 2\frac{\beta}{\mathcal{W}}} \qquad (2.16)$$

where $\wp^+$ and $\wp^-$ are solutions for loading and unloading, respectively. In loading, $\wp^+(r,\theta)$ is calculated by considering the parameters with (+) sign in (2.15), and in unloading, $\wp^-(r,\theta)$ is calculated by considering the parameters with (−) sign in (2.15). The parameters and coefficients in (2.16) are given by

$$G_1 = -\frac{3}{8}(F_1)^2 + F_2, G_1^* = F_2^*, G_2 = \frac{1}{8}(F_1)^3 - \frac{1}{2}F_1F_2 + F_3, G_2^* = -\frac{1}{2}F_1F_2^* + F_3^*,$$

$$G_3 = -\frac{3}{256}(F_1)^4 + \frac{1}{16}(F_1)^2F_2 - \frac{1}{4}F_1F_3 + F_4, G_3^1 = \frac{1}{16}(F_1)^2F_2^* - \frac{1}{4}F_1F_3^* + F_4^*,$$

$$K_1 = -\frac{1}{12}(G_1)^2 - G_3, K_1^* = -\frac{1}{6}G_1G_1^* - G_3^1, K_1^{\diamond} = -\frac{1}{12}(G_1^*)^2,$$

$$K_2 = -\frac{1}{108}(G_1)^3 + \frac{1}{3}G_1G_3 - \frac{1}{8}(G_2)^2, K_2^* = -\frac{1}{36}(G_1)^2G_1^* + \frac{1}{3}G_1G_3^1 + \frac{1}{3}G_3G_1^* - \frac{1}{4}G_2G_2^*,$$

$$K_2^{\diamond} = -\frac{1}{36}G_1G_1^* + \frac{1}{3}G_1^*G_3^1 - \frac{1}{8}(G_2^*)^2, K_2^{\circ} = -\frac{1}{108}(G_1^*)^3 \qquad (2.17)$$

and

$$\mathcal{R} = -\frac{1}{2}K_2 - \frac{1}{2}K_2^*r\theta - \frac{1}{2}K_1^{\diamond}r^2\theta^2 - \frac{1}{2}K_2^{\circ}r^3\theta^3$$

$$+ \left[\frac{1}{4}(K_2)^2 + \frac{1}{2}K_2K_2^*r\theta + \frac{1}{2}K_2K_2^{\diamond}r^2\theta^2 + \frac{1}{2}K_2K_2^{\circ}r^3\theta^3 + \frac{1}{4}(K_2^*)^2r^2\theta^2\right.$$

$$+ \frac{1}{2}K_2^*r^3\theta^3K_2^{\diamond} + \frac{1}{2}K_2^*K_2^{\circ}r^4\theta^4 + \frac{1}{4}(K_2^{\diamond})^2r^4\theta^4 + \frac{1}{2}K_2^{\diamond}K_2^{\circ}r^5\theta^5 + \frac{1}{4}(K_2^{\circ})^2r^6\theta^6$$

$$+ \frac{1}{27}(K_1)^3 + \frac{1}{9}(K_1)^2K_1^*r\theta + \frac{1}{9}(K_1)^2K_1^{\diamond}r^2\theta^2 + \frac{1}{9}K_1(K_1^*)^2r^2\theta^2 + \frac{2}{9}K_1K_1^*K_1^{\diamond}r^3\theta^3$$

$$+ \left. \frac{1}{9}K_1(K_1^{\diamond})^2r^4\theta^4 + \frac{1}{27}(K_1^*)^3r^3\theta^3 + \frac{1}{9}(K_1^*)^2K_1^{\diamond}r^4\theta^4 + \frac{1}{9}K_1^*(K_1^{\diamond})^2r^5\theta^5 + \frac{1}{27}(K_1^{\diamond})^3r^6\theta^6\right]^{1/2},$$

$$\mathcal{U} = (\mathcal{R})^{1/3}, \quad \mathcal{Q} = K_2 + K_2^*r\theta + K_2^{\diamond}r^2\theta^2 + K_2^{\circ}r^3\theta^3, \quad \alpha = G_1 + G_1^*r\theta, \quad \beta = G_2 + G_2^*r\theta,$$

$$\mathcal{P} = K_1 + K_1^*r\theta + K_1^{\diamond}r^2\theta^2, \quad \mathcal{Y} = -\frac{5}{6}\alpha + \mathcal{U} - \frac{1}{3}\frac{\mathcal{P}}{\mathcal{U}}, \quad \mathcal{W} = \sqrt{\alpha + 2\mathcal{Y}} \qquad (2.18)$$

---

[1] In 1540, Lodovico Ferrari found the solution of quartic equation by reducing it to a cubic equation. However, because the solution for cubic equations was not available at the time, his solution was not published. Four years later, Ferrari's teacher, Gerolamo Cardano, published the solution of both quartic and cubic equations in his book *Ars Magna* [47].

This analytical solution can be used to investigate the influence of several different parameters, for example, temperature and material properties on the torsional response of SMA circular bars [48]. Also, this solution can be developed to analyze SMA helical springs as explained in the next section.

2.1.3 SMA Helical Springs

The theoretical solution for analyzing the torsion of SMA bars is improved for developing a closed-form solution for studying SMA helical springs. Several applications with SMA helical spring actuators have been introduced recently. In this section, we focus on studying the pseudoelastic response of SMA helical springs under axial force analytically and numerically. In the analytical solution, two different approximations are considered. In the first approximation, both the curvature and pitch effects are assumed to be negligible. This is the case for helical springs with large ratios of mean coil radius to the cross-sectional radius (spring index) and small pitch angles. Using this assumption, analysis of the helical spring is reduced to that of the pure torsion of a straight bar with circular cross section, and the solution of the previous section is used (denoted as straight bar torsion model or SBTM).

Based on the SBTM assumption, the spring is considered as a straight bar of length $l = 2\pi N R_m$, where N is the number of active coils; the total angular deflection of one end of the bar with respect to the other end is given by $\Theta = 2\pi N R_m \theta$. Because the effective moment arm of the axial load F is equal to R_m, the deflection of the spring at the end point is given by

$$\delta = \Theta R_m = 2\pi N R_m^2 \theta \qquad (2.19)$$

For any deflection of the spring ends, (2.19) is used for finding the twist angle per unit length θ. Substituting this value in (2.16), the shear stress distribution in the cross section is calculated. Having the shear stress distribution, the resultant torque in the cross section, $\mathbb{T}$, is obtained by integrating over the surface, and the axial force corresponding to the assumed end displacement is calculated by dividing the torque by the coil mean radius: $F = \mathbb{T}^{\pm}/R_m$. In this approximation, the effect of direct shear force on the cross section is ignored along with the curvature and pitch effects. In the next step, the curvature effect is included, and the SMA helical spring is analyzed using the exact solution presented for torsion of curved SMA bars. In this refined solution, the effect of the direct shear force is also considered.

It is known that due to the curvature effect, the shear stress distribution in the cross section is not axisymmetric. The following solution is applicable for

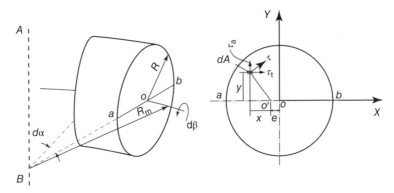

Figure 2.1   Torsion of a curved bar

SMA helical springs with large spring indices but small pitch angles. This includes nearly all the practical helical springs [49]. This solution method is denoted by the curved bar torsion model (CBTM) throughout the text.

First, we consider the pure torsion of an SMA curved bar. A slice of such a bar is shown in Figure 2.1. By applying a torque, the two faces of this cross section will rotate with respect to each other by an angle $d\beta$. Since the initial length of the filament passing through the points a and b are not the same, the strain distribution in the cross section is not axisymmetric. This will result in a nonaxisymmetric stress distribution with a larger value for the material points near the axis AB. The nonaxisymmetric shear stress distribution in the cross section can be decomposed into an axial component τ_a and a transverse component τ_t as shown in Figure 2.1. If we assume that the cross section is rotating about an axis passing through the center o (at the center of rotation, the shear stress is zero), and considering the fact that the shear stress at b is larger than the stress at a, such a distribution cannot be in equilibrium; there will be a torque in the cross section [49]. So, it can be concluded that for a curved bar under torsion, the zero shear stress point or the center of rotation does not coincide with the cross-sectional center. Considering the symmetry conditions in the cross section, the forces caused by the transverse shear stress component τ_t are in equilibrium when the rotation center is anywhere on the axis ab. The rotation center o' is shown in Figure 2.1. The distance e for finding the location of the rotation center is found by the method presented in the sequel.

In the coordinate system xy with the origin o', when the slice sides rotate by the amount $d\beta$ with respect to each other, the relative movement of the ends of any filament corresponding to dA in Figure 2.1 is $\sqrt{x^2 + y^2}\, d\beta$. Considering the fact that the length of this filament in the undeformed configuration is $(R_m - e - x)d\alpha$, the shear strain corresponding to this point in the cross section is

$$\gamma = \frac{\sqrt{x^2 + y^2}}{R_m - e - x}\frac{d\beta}{d\alpha} \qquad (2.20)$$

where the shear strain γ is along the direction of τ in Figure 2.1. The geometrical parameters in (2.20) are shown in Figure 2.1. Now, the expression given for the shear strain in (2.20) should be replaced with the expression $r\theta$ (i.e., the strain in the straight bar case) in (2.14). The new quartic equation obtained by this substitution is solved to find the explicit expression for the shear stress in the regions with phase transformation. We denote the shear stress in this case by $\tau = \hat{\wp}^{\pm}(x, y, e, d\beta/d\alpha)$, where the explicit expression for shear stress is similar to that obtained for straight bar by replacing the parameter $r\theta$ with the shear strain of a curved bar in (2.20). For a curved bar under pure torsion, the resultant force in the cross section should be zero. Due to symmetry of the transverse shear stress about the axis ab, the forces caused by this stress component are in static equilibrium. The static equilibrium for the forces caused by the axial component of the shear stress is expressed by

$$\int_A \tau_a dA = \int_{A_1} \frac{G_A x}{R_m - e - x}\frac{d\beta}{d\alpha} dA_1 + \int_{A_2} \hat{\wp}^{\pm}(x, y, e, d\beta/d\alpha)\frac{x}{\sqrt{x^2 + y^2}} dA_2$$

$$+ \int_{A_3} \left(\tau_a^f + \frac{G_M x}{R_m - e - x}\frac{d\beta - d\beta^f}{d\alpha}\right) dA_3 = 0 \qquad (2.21)$$

where A_1 is that portion of the section that has not experienced the phase transformation. The portion of the cross section with phase transformation is denoted by A_2 and the parts in which the phase transformation has been completed by A_3. For each material point with completed phase transformation ($\xi = 1$), the parameter τ_a^f is the axial component of shear stress corresponding to $d\beta^f$ that is the twist angle of phase transformation completion for that material point. Now, the only unknown parameter in (2.21) is the position of the rotation center e. In contrast with the elastic torsion of a bar, in the case of an SMA curved bar, the second and third integrals in (2.21) cannot be calculated analytically, but numerical integration method can be used.

For analyzing SMA helical springs based on the pure torsion of an SMA curved bar, a minor correction is needed to take into account the direct shear force in the cross section. In an SMA helical spring, a direct shear force F and a torque R_mF are acting in each cross section. The equilibrium equations in the cross section in this case read

$$\int_A \tau_a dA = F, \text{ and } \int_A \tau_a x dA + \int_A \tau_t y dA = R_m F \tag{2.22}$$

where in the most general case, the cross section is divided into three regions similar to the straight bar case. For analyzing the SMA spring using the curved bar theory, a predefined displacement is considered for the spring ends. The total rotation of the spring ends with respect to each other is calculated by $\hat{\Theta} = \delta/R_m$. The twist angle per curvature angle is given by

$$\frac{d\beta}{d\alpha} = \frac{\delta}{2\pi R_m N} \tag{2.23}$$

The shear stress is calculated by the above given method. Since the axial force F is unknown, in contrast with the pure torsion case, the parameter e cannot be obtained directly by solving (2.22). Hence, first, an initial value is considered for e. Using this value, the axial force F is calculated using both expressions in (2.22). A trial and error method is then used to find the value of e for which the difference of the axial forces calculated from the two conditions in (2.22) is smaller than a tolerance (i.e., 1N).

2.1.4 Superelastic Bending of SMA Beams

Bending is one of the most important loading conditions in designing several SMA actuators. The goal of this section is obtaining a closed-form solution for analyzing the bending of SMA beams using the mentioned macroscopic phenomenological constitutive framework. In this section, two different transformation functions will be considered: a J_2-based model with symmetric tension–compression response and a J_2–I_1-based model for considering the tension–compression asymmetry that is observed in experiments. The constitutive equations are reduced to an appropriate form for studying the pseudoelastic bending response of SMAs. Closed-form expressions are given for the stress and martensitic volume fraction distributions in the cross section, and the bending moment–curvature relation is obtained analytically. Both circular and rectangular cross sections can be studied by this method.

By appropriate selection of the function $\tilde{\Phi}(\sigma)$ in the transformation criteria (2.6), different material responses observed in experiments can be modeled. There are numerous selections for the transformation function of SMAs in the literature based on J_2 [41], J_2-J_3 [49], J_2-I_1 [51], and $J_2-J_3-I_1$ [44]. The models with a transformation function based on a J_2 invariant are the simplest and the best choice for our purposes of seeking a closed-form solution. However, by developing the constitutive equations based on J_2 invariant, although the majority of the SMA experimentally observed responses are modeled with good accuracy, the tension–compression asymmetry (which plays an important rule in bending) cannot be modeled. We will use a J_2-based model and also modify it by using a J_2-I_1 model for taking into account the tension–compression asymmetry. The function $\tilde{\Phi}(\sigma)$ for a J_2-invariant-based model is given by

$$\tilde{\Phi}(\sigma) = \aleph\sqrt{3J_2} = \aleph\sqrt{\frac{3}{2}\sigma':\sigma'} \qquad (2.24)$$

where $\aleph$ is a material constant corresponding to the maximum transformation strain during forward phase transformation in tension or compression and σ' is the deviatoric stress.

The selection of transformation function based on the J_2 invariant results into a symmetric response in tension and compression. However, it is experimentally well known that single crystal and polycrystalline shape memory alloys have a nonsymmetric tension–compression response [52–54]. There have been numerous efforts in the literature for better understanding the origins of this secondary effect in SMAs and introducing appropriate constitutive relations capable of modeling this effect [55–57]. Most of the existing constitutive relations for modeling the tension–compression asymmetry are appropriate only for numerical simulations and not for closed-form solutions because of their complexity. We use the J_2-I_1-based transformation function that enables the constitutive relations to model the tension–compression asymmetry besides relative simplicity compared to the other models [44, 51]. The function $\tilde{\Phi}(\sigma)$ for this model is given by

$$\tilde{\Phi}(\sigma) = \eta\sqrt{3J_2} + \omega I_1 = \eta\sqrt{\frac{3}{2}\sigma':\sigma'} + \omega\,\mathrm{tr}(\sigma) \qquad (2.25)$$

where η and ω are material constants related to the maximum transformation strains during forward phase transformation in tension and compression.

In the special case that the only nonzero stress component is the normal stress σ_x, the constitutive relations can be simplified for obtaining closed-form expressions for solving the superelastic bending of SMA beams. For pure bending, the explicit expressions for the martensitic volume fraction in direct and inverse phase transformation for J_2 and J_2–I_1 models in pure bending are given by [58]

$$\xi^{\pm} = \frac{1}{\rho b^{\pm}}\left\{ \aleph|\sigma_x| + \frac{1}{2}\sigma_x^2\Delta S_{11} + \rho\Delta s_0(T-T^{\pm})\right\} \tag{2.26}$$

for the J_2-based model and

$$\xi^{\pm} = \frac{1}{\rho b^{\pm}}\left\{ \eta|\sigma_x| + \omega\sigma_x + \frac{1}{2}\sigma_x^2\Delta S_{11} + \rho\Delta s_0(T-T^{\pm})\right\} \tag{2.27}$$

for the J_2–I_1-based model. Also, the constitutive equation can be reduced to the following one-dimensional form for the pure bending:

$$\epsilon_x = \left(S_{11}^A + \xi\Delta S_{11}\right)\sigma_x + \alpha_A(T-T_0) + (\ell_c\hat{\eta} + \hat{\omega})\xi \tag{2.28}$$

where $S_{11}^A = 1/E_A$, $\Delta S_{11} = 1/E_M - 1/E_A$ (E_A and E_M are the elastic muduli of austenite and martensite, respectively). Substituting the martensitic volume fractions (2.26), (2.27), and $\epsilon_x = -\kappa y$, where κ is the axis curvature and y is the distance from the neutral axis into (2.28), the stress–strain relation can be written as the following cubic equation:

$$\sigma_x^3 + a\sigma_x^2 + b\sigma_x + \tilde{c} + \tilde{\kappa}y = 0 \tag{2.29}$$

where a, b, $\tilde{c}$, and $\tilde{\kappa}$ are constants given by

$$a = \frac{3(\ell_c\hat{\eta} + \hat{\omega})}{\Delta S_{11}}, \tilde{\kappa} = \frac{2\kappa\rho b^{\pm}}{\Delta S_{11}^2},$$

$$b = \frac{2\rho\Delta s_0(T-T^{\pm})}{\Delta S_{11}} + \frac{2(\ell_c\hat{\eta} + \hat{\omega})^2 + 2\rho b^{\pm}S_{11}^A}{\Delta S_{11}^2}, \tag{2.30}$$

$$\tilde{c} = \frac{2(\ell_c\hat{\eta} + \hat{\omega})\rho\Delta s_0(T-T^{\pm}) + 2\rho b^{\pm}\alpha_A(T-T_0)}{\Delta S_{11}^2}$$

The cubic equation (2.29) is solved for σ_x as a function of temperature and strain. The acceptable roots for the SMA material in tension and compression are

$$\sigma_t = \frac{1}{6}(A - 108\tilde{\kappa}y + \mathcal{P})^{1/3} - \frac{2b - 2a^2/3}{(A - 108\tilde{\kappa}y + \mathcal{P})^{1/3}} - \frac{a}{3}, \tag{2.31}$$

$$\sigma_c = \frac{-1}{12}(A - 108\tilde{\kappa}y + \mathcal{P})^{1/3} + \frac{b - a^2/3}{(A - 108\tilde{\kappa}y + \mathcal{P})^{1/3}} - \frac{a}{3}$$
$$- \frac{\sqrt{3}}{2}i\left[\frac{1}{6}(A - 108\tilde{\kappa}y + \mathcal{P})^{1/3} + \frac{2b - 2a^2/3}{(A - 108\tilde{\kappa}y + \mathcal{P})^{1/3}}\right] \tag{2.32}$$

where $A = 36ab - 108\tilde{c} - 8a^3$, $B = 162\tilde{c} - 54ab + 12a^3$, $C = 12b^3 - 3a^2b^2 - 54ab\tilde{c} + 12a^3\tilde{c} + 81\tilde{c}^2$, and $\mathcal{P} = 12\sqrt{81\tilde{\kappa}^2y^2 + B\tilde{\kappa}y + C}$. The explicit expressions in (2.31) give the exact value of stress. These expressions can be further simplified by implementing some assumptions on the material properties. The trigonometric form of the roots of the cubic equation (2.29) can also be used for obtaining a simplified stress–curvature relation as [58, 59]

$$\sigma = \left[\ell_c\left(1 - \frac{1}{2}\beta^2\right)\cos\varphi + \beta\sin\varphi\right]g - \frac{a}{3} \tag{2.33}$$

where $\theta = 12\sqrt{-81\tilde{\kappa}^2y^2 - B\tilde{\kappa}y - C}/(A - 108\tilde{\kappa}y)$, $g = \frac{1}{3}(A^2 - 144C)^{1/6}$, $\varphi = \pi/6$, and $\beta = 1/(3\theta)$. The bending moment–curvature relationship for an SMA beam with an arbitrary cross section is given by $M = \int_\Omega y\sigma(y)dA$, where M is the bending moment, y is the distance from the neutral axis, and Ω is the cross section. In the most general case, the cross section is divided into three regions: an elastic core in which the phase transformation has not started, a middle part with phase transformation, and the outer part in which the material is fully transformed to martensite. In order to calculate the total bending moment, the bending moment in each part should be found and summed in the whole cross section. The most complicated section to be solved is the middle part with active phase transformation. However, the bending moment in this section can be calculated explicitly by using the obtained stress distributions (2.31). The boundaries of the region in which the phase transformation occurs can be found by considering the limit values of $\xi^+ = 0$ and $\xi^+ = 1$ at these boundaries. The upper and lower heights of the phase transformation region in the cross section are then obtained as

$$
y_{1t} = \frac{\sigma_s|\ell_c = 1}{\kappa E^A}, \; y_{2t} = \frac{(\hat{\eta} + \hat{\omega})E^M + \sigma_f|\ell_c = 1}{\kappa E^M},
$$

$$
y_{1c} = \frac{\sigma_s|\ell_c = -1}{\kappa E^A}, \; y_{2c} = \frac{(-\hat{\eta} + \hat{\omega})E^M + \sigma_f|\ell_c = -1}{\kappa E^M}
$$

(2.34)

where y_1 and y_2 are the lower and upper bonds of the region with phase transformation, respectively. Now, the bending moment at each cross section along the length of the superelastic beam can be calculated by $M = \int_\Omega y\sigma(y)dA$ which is given by

$$
\begin{aligned}
M = &-\frac{1}{3}E^A \kappa w \left(y_{1c}^3 - y_{1t}^3\right) + \left(\mathscr{I}|_{y=y_{2c}} - \mathscr{I}|_{y=y_{1c}}\right) \\
&+ E^M w \left[\frac{1}{3}\kappa\left(\frac{h_c^3}{8} - y_{2c}^3\right) - H^c\left(\frac{h_c^2}{4} - y_{2c}^2\right)\right] + \left(\mathscr{I}|_{y=y_{1t}} - \mathscr{I}|_{y=y_{2t}}\right) \\
&+ E^M w \left[\frac{1}{3}\kappa\left(y_{2t}^3 - \frac{h_t^3}{8}\right) - H^c\left(y_{2t}^2 - \frac{h_t^2}{4}\right)\right]
\end{aligned}
$$

(2.35)

The parameter $\mathscr{I}$ in (2.35) corresponds to the integral $\int_\Omega y\sigma(y)dA$ at the region with active phase transformation, which for a rectangular cross section is given by

$$
\begin{aligned}
\mathscr{I} = \int y\sigma(y)wdy = &\frac{-wg\sqrt{R}(-54\tilde{\kappa}y + A + B)\sin\varphi}{2916\tilde{\kappa}^2} \\
&-\frac{w\ell_{cg}y(3A + 2B)\cos\varphi}{2916\tilde{\kappa}} + \frac{19}{36}wy^2\ell_{cg}\cos\varphi - \frac{1}{6}wy^2a \\
&-\frac{wg\tan^{-1}(S)(B^2 - 108C + AB)\sin\varphi}{2^3 3^8 \tilde{\kappa}^2} \\
&+\frac{w\ell_{cg}\ln(-R)(-1296C + 9A^2 + 24AB + 16B^2)\cos\varphi}{2^6 3^{10}\tilde{\kappa}^2} \\
&+ w\ell_{cg}\tanh^{-1}(Q)\cos\varphi \\
&\left(\frac{-3888C(A + B) + 9A^2B + 24AB^2 + 16B^3}{2^5 3^{10}\tilde{\kappa}\sqrt{B^2\tilde{\kappa}^2 - 324\tilde{\kappa}^2C}}\right) + \mathbb{C}_1
\end{aligned}
$$

(2.36)

Table 2.1 SMA material parameters

Material constants	$Ni_{50}Ti_{50}$[60]
E^A	72.0×10^9Pa
E^M	30.0×10^9Pa
$\nu^A = \nu^M$	0.42
$\rho c^A = \rho c^M$	2.6×10^6 J/(m³K)
H^t	0.05
H^c	-0.035
$(d\sigma/dT)_t^A$	8.4×10^6 J/(m³K)
$\rho \Delta s_0 = -H^t(d\alpha/dT)_t^A$	-0.42×10^6 J/(m³K)
A_f	281.6 K
A_s	272.7 K
M_s	254.9 K
M_f	238.8 K

where w is the cross-sectional width, $\mathbb{C}_1$ is a constant of integration, and

$$R = -81\tilde{k}^2 y^2 - Bky - C, \quad Q = \frac{162\tilde{k}^2 y + B\tilde{k}}{\sqrt{B^2\tilde{k}^2 - 324\tilde{k}^2 C}},$$

$$S = \frac{9\tilde{k}}{\sqrt{R}}\left(y + \frac{1}{162}\frac{B}{\tilde{k}}\right) \tag{2.37}$$

Having the bending moment–curvature relation (2.35) completes the closed-form solution which is capable of predicting the load–deflection response and the stress and martensitic volume fraction distributions accurately. As an example, consider an SMA cantilever with length $L = 10$ cm. The rectangular cross section has a height of $h = 1$ cm and width $w = 1.5$ mm. The J_2-based model is used and the material properties for $Ni_{50}Ti_{50}$ are given in Table 2.1 [60].

The analytical results obtained from the present formulation are also compared with those of a three-dimensional finite element simulation. The temperature is $T = T_0 = 300$ K, and an isothermal loading–unloading process is assumed. The superelastic cantilever is subjected to a transverse tip load of 200 N. The stress distribution at the end of the loading phase is shown in Figure 2.2a at the clamped edge. The contour plots of the normal stress distribution at the end of the loading phase obtained from the present closed-form solution is also shown in Figure 2.2b. The martensitic volume fraction at the

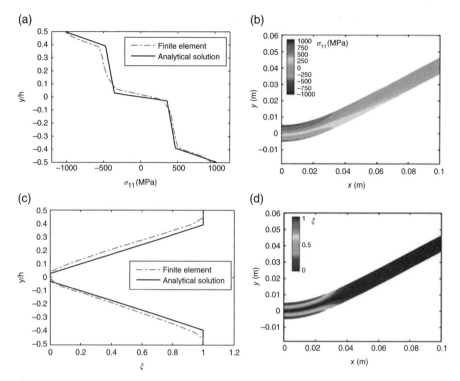

Figure 2.2 Comparison of the finite element and analytical results for (a) normal stress and (c) martensitic volume fraction distribution at the clamped edge of an SMA beam with rectangular cross section subjected to bending. Contour plots of the normal stress and the martensitic volume fraction distributions obtained by the analytical solution are shown in (b) and (d), respectively

clamped edge and the contour plot of the martensitic volume fraction distribution are shown in Figure 2.2c and d, respectively. As it is shown, the closed-form solution calculates both the stress and martensitic volume fractions accurately.

As shown in Figure 2.2, the core remains austenite without phase transformation. By considering pure bending, the stress around the neutral axis is zero, and there is always an austenite core without phase transformation even for large deflections. By considering the shear effect in bending, the stress at the core is nonzero which may cause phase transformation at the core as well. However, except for very thick beams, the pure bending theory gives accurate results, and this is reflected in comparison of the results with the finite element solution that considers the shear effect. It is worth noting that in the closed-form solution, the nonlinear geometry effects and the displacement along the beam axis direction are ignored, while the numerical simulations show a minor deflection along the axis due to geometric nonlinearities. As shown in Figure 2.2a and c, the present method gives accurate results even for the large deflection chosen in these case studies.

2.1.5 Thermomechanical Coupling in the Response of SMAs

The martensitic phase transformation in SMAs is associated with generation or absorption of latent heat in forward (austenite to martensite) and reverse (martensite to austenite) transformations. This has been shown in many experiments, and the heat of transformation and the associate temperatures for the start and end of forward/reverse martensitic transformation can be determined by differential scanning calorimeter or DSC [61, 62]. In the majority of the previous works on SMAs in which loading is considered quasistatic, it is assumed that the material is exchanging the phase transformation-induced latent heat with the ambient such that the SMA device is always isothermal and in a temperature identical with the ambient during loading and unloading. However, definition of quasistatic loading that guarantees an isothermal process is not absolute; it is affected by a number of parameters, for example, the ambient condition and size of the structure. In other words, a very slow loading rate that can be considered a quasistatic loading for an SMA wire with a small diameter may be far from being quasistatic and isothermal for a bar with larger diameters. This size effect phenomenon has been also reported previously in some experiments [63, 64]. In some of the previously reported works in the literature, the effect of this latent heat and its coupling with mechanical response of SMAs was considered along with some simplifying assumptions.

The thermomechanical coupling in SMAs can be studied analytically based on the closed-form solutions for various devices subjected to different loading conditions. In each case, the closed-form solutions for the adiabatic case are used, and the energy equation in a rate form is coupled to the constitutive equations. In addition to the theoretical studies, the phase transformation-induced latent heat in SMAs can also be studied experimentally by temperature and strain measurement equipments. These experimental results are only applicable for measuring the temperature at the surface. The measured temperatures by this method can be used for calibrating the theoretical models

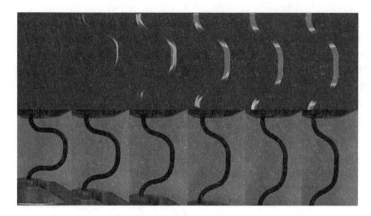

Figure 2.3 Experiments on superelastic deformation of a U-shaped SMA device. (Top) the temperature distribution measured by the infrared camera and (bottom) the deformed shape

at the surface, and then the theoretical framework is capable of predicting the temperature changes inside the material, which is impractical to be measured in the experiments.

Thermovision camera can be used to record the reflection of infrared radiation from the specimen for finding its temperature variations at the surface during various loading conditions. As an example, the temperature distribution obtained from the infrared camera for an SMA U-shaped device (with energy absorption applications) is shown in Figure 2.3. The deformation of the device is also shown in this figure. It is observed that the phase transformation latent heat is affecting the temperature at different regions and the material is not in isothermal conditions, despite the relative slow loading rate that was used in this experiment.

In order to obtain the coupled thermomechanical governing equations for SMAs in the phenomenological models, we start from the first law of thermodynamics in local form

$$\rho \dot{u} = \sigma : \dot{e} - \mathrm{div}\, q + \rho \hat{g} \qquad (2.38)$$

where ρ is the mass density, u is the internal energy per unit mass, and the parameters q and $\hat{g}$ are the heat flux and internal heat generation, respectively. The dot symbol on a quantity ($\cdot$) represents time derivative of the quantity. The dissipation inequality reads

$$\rho \dot{s} + \frac{1}{T}\mathrm{div}\, q + \frac{\rho \hat{g}}{T} \geq 0 \qquad (2.39)$$

where s is the entropy per unit mass. Substituting the Gibbs free energy into the dissipation inequality, another form of the second law of thermodynamics is obtained as

$$-\rho \dot{G} - \dot{\boldsymbol{\sigma}} : \epsilon - \rho s \dot{T} \geq 0 \tag{2.40}$$

After some mathematical manipulations, the first law of thermodynamics can be rewritten as [45]

$$T\alpha : \dot{\boldsymbol{\sigma}} + \rho c \dot{T} + \left[-\mathcal{P} + T\Delta\alpha : \sigma - \rho\Delta c T \ln\left(\frac{T}{T_0}\right) + \rho\Delta s_0 T \right] \dot{\xi} = -\operatorname{div} q + \rho \hat{g} \tag{2.41}$$

Also, the martensitic volume fraction rate can be expressed as [45]

$$\dot{\xi} = -\frac{(\boldsymbol{\Gamma} + \Delta\mathbb{S} : \boldsymbol{\sigma}) : \dot{\boldsymbol{\sigma}} + \rho\Delta s_0 \dot{T}}{\mathfrak{D}^{\pm}} \tag{2.42}$$

where $\mathfrak{D}^{+} = \rho\Delta s_0(M_s - M_f)$ for the forward phase transformation $(\dot{\xi} > 0)$ and $\mathfrak{D}^{-} = \rho\Delta s_0(A_s - A_f)$ for reverse phase transformation $(\dot{\xi} < 0)$. Substituting (2.42) into (2.66) and assuming $\Delta\alpha = \Delta c = 0$—valid for almost all practical SMA alloys—the following expression is obtained:

$$[T\alpha - \mathcal{F}_1(\sigma, T)] : \dot{\boldsymbol{\sigma}} + [\rho c - \mathcal{F}_2(T)]\dot{T} = -\operatorname{div} q + \rho \hat{g} \tag{2.43}$$

where

$$\mathcal{F}_1(\sigma, T) = \frac{1}{\mathfrak{D}^{\pm}}(\boldsymbol{\Gamma} + \Delta\mathbb{S} : \boldsymbol{\sigma})(\mp Y + \rho\Delta s_0 T), \mathcal{F}_2(T) = \frac{\rho\Delta s_0}{\mathfrak{D}^{\pm}}(\mp Y + \rho\Delta s_0 T) \tag{2.44}$$

In (2.44), (+) is used for forward phase transformation, and (−) is used for the reverse transformation. Equation 2.43 is one of the two coupled relations for describing the thermomechanical response of SMAs. The second relation is the constitutive Equation 2.3. A similar procedure can be followed for obtaining the coupled equations corresponding to the pure torsion of SMA bars [65].

The obtained coupled equations can be discretized for wires and bars with circular cross sections using an explicit finite difference method for several loading conditions including tension [45], torsion [65], and combined loadings [66]. The discretized form of convection boundary conditions is also derived.

For modeling SMA wires and bars operating in still air and exposed to air or fluid flow with a known speed, free and forced convection coefficients are calculated for slender wires and thick cylindrical bars in air and fluid using the experimental and analytical formulas in the literature.

By this analytical model, it is shown that a loading condition of quasistatic or dynamic strongly depends on the ambient conditions and the specimen size besides the loading rate. The temperature distribution nonuniformity inside a device can be considered as a measure for the validity of using quasistatic and isothermal assumptions. For example, consider a circular bar subjected to axial loading. The initial and ambient temperatures are assumed to be $T_0 = T_\infty = 300$ K. For studying the nonuniformity in stress and temperature distributions inside the bar, the difference between the value of these parameters at the center and surface of the bar is nondimensionalized by dividing by the value of the corresponding parameter at the surface. The maximum temperature and stress gradients versus the convection coefficient and loading rate are shown in Figure 2.4. A bar with $d = 5$ cm and material properties similar to the previous case study is considered, and the range of convection coefficient is chosen to cover the free and forced convection of air and water flow on the bar (see [45] for more examples of SMA bars operating in water). As it is shown in this figure for all the loading rates, both the temperature and stress gradients increase for larger convection coefficients. However, for the slow loading rates $\tau > 300$ s, the gradients are negligible even for large convection coefficients $h_\infty > 1000$ W/m^{2}K, since the material has enough time to exchange heat with the ambient. Also, it is seen that the temperature and stress gradients

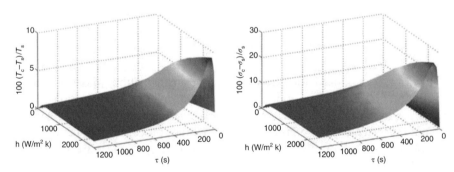

Figure 2.4   The maximum temperature gradient (left) and the maximum stress gradient (right) versus the loading rate and convection coefficient. The subscripts s and c represent the values measured at the surface and center of bar, respectively. The diameter of bar is d = 5 cm

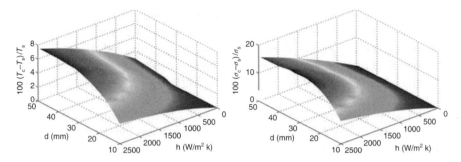

Figure 2.5   The maximum temperature gradient (left) and the maximum stress gradient (right) versus the diameter and convection coefficient. The subscripts s and c represent the values measured at the surface and center of bar, respectively. The total loading–unloading time is $\tau = 10$ s

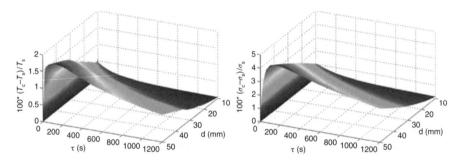

Figure 2.6   The maximum temperature gradient (left) and the maximum stress gradient (right) versus the diameter and loading rate. The subscripts s and c represent the values measured at the surface and center of bar, respectively. The convection coefficient is $h_\infty = 200$ W/m²K

are small for very fast and very slow loading rates and the nonuniformity in both the variables peaks at an intermediate loading rate ($\tau \simeq 140$ s). A similar trend is seen for all the convection coefficients, although the nonuniformity is more intensive for larger convection coefficients.

The effect of changing bar diameter and boundary conditions on the temperature and stress nonuniformity is studied as shown in Figure 2.5 (the total loading–unloading time is $\tau = 10$ s). As it is shown in Figure 2.5, the temperature and stress nonuniformities are more pronounced for larger diameters and convection coefficients. The effect of changing bar diameter and the loading rate on the gradients is studied as depicted in Figure 2.6 (the convection coefficient is $h_\infty = 200$ W/m^{2}K). As it is shown in Figure 2.5, the temperature and stress nonuniformities are more pronounced for larger diameters. The effect of loading rate on the gradients is more complicated. The maximum gradient for both the temperature and stress distributions happens for an intermediate loading rate of $\tau \simeq 120$ and decreases for very slow and fast loadings.

The results presented in Figures 2.4–2.6 clearly describe the complicated effect of size, ambient conditions, and loading rate on the coupled thermomechanical response of SMA bars. These figures can be used by a designer to decide whether a coupled thermomechanical formulation with considering the heat flux in the cross section is necessary or using simpler lumped models is enough. It is worth mentioning that although for the uniaxial loading of bars and wires the simpler models assuming lumped temperature in the cross section can be used with an error, there are numerous cases for which the present formulation is the only analysis option. An example is torsion or combined loading of circular SMA bars for which shear stress has a complicated nonuniform distribution in the cross section [47, 66]. It would be incorrect to consider a lumped temperature in the cross section for torsion problems.

2.2 Micromechanical Modeling

Recent applications of SMAs as actuators in MEMS motivate us to develop micromechanical-based frameworks to study the response of SMA polycrystals with a limited number of grains. This micromechanical framework can be used to study some important phenomena in the response of SMAs including the rule of texture, the effect of grain size and shape, the intergranular stresses, and the texture development during phase transformation in various loading conditions. For modeling polycrystalline SMAs in this scale, a multivariant micromechanical model will be used for incorporating single crystal constitutive relationships. The model is based on considering 24 possible twinning systems in martensitic transformation and calculating the total transformation strain by summing the transformation strain of all the activated twinning systems. The polycrystalline texture measured by the X-ray diffraction is given as the input to the micromechanical model, and different textures can be studied by this model. For modeling the polycrystalline structure, we will use a Voronoi tessellation. Voronoi tessellations are widely accepted to model polycrystalline aggregates as they provide a realistic approximation of the actual microstructure of nonuniform grain shapes. Voronoi cells are constructed from a set of randomly positioned points (called the generators or Poisson points) in the given domain. Each Voronoi cell is the set of all points in the given set whose distance to the corresponding generator is not greater than their distance to the other generators. The obtained constitutive equations are used for developing a three-dimensional finite element model.

2.2.1 The Micromechanical Framework

The micromechanical framework that was introduced in Ref. [55] is adopted and developed in this section. This model was initially applied to study the role of texture in tension-compression asymmetry in polycrystalline NiTi [1], also has been implemented in a three-dimensional finite element formulation for studying the cyclic thermomechanical behavior of polycrystalline pseudoelastic shape memory alloys [67]. The micromechanical-based model considers the stress-free transformation strain corresponding the nth martensite variant as

$$\hat{\epsilon}^n_{ij} = \frac{1}{2}g\left(l^n_i d^n_j + d^n_i l^n_j\right) \tag{2.45}$$

where d is the transformation direction, l is the habit plane normal, and g is the transformation magnitude. We consider the B2 $\rightarrow$ B19′ martensitic transformation for NiTi. By considering the most dominant experimentally observed twin modes (type II-1), the number of martensite variant pairs for the transformation is $1 \leq n \leq 24$, and the habit plane and transformation direction components are obtained from the crystallographic data (see Table 1 in [1]). The transformation magnitude, which represents the magnitude of deformation of a matrix austenite volume element into a martensitic one, for this case is $g = 0.1308$. The total transformation strain in a single grain is the sum of transformation strain from all martensite variants. By defining a martensitic volume fraction corresponding to each variant, the total transformation strain is given by

$$\epsilon^t_{ij} = \sum_{n=1}^{24} \hat{\epsilon}^n_{ij}\xi^n \tag{2.46}$$

where ξ^n is the volume fraction of the nth martensite variant. The total martensite volume fraction is $\xi = \sum_{n=1}^{24}\xi^n$. The total volume fraction and the volume fraction of each variant is always between 0 and 1. For a single crystal, the Gibbs free energy is given by

$$G(\sigma, T, \xi^n) = \frac{1}{2}\sigma : \mathbb{S} : \sigma + \sigma : \sum_{n=1}^{24}\hat{e}^n\xi^n - \beta(T - T_0)\sum_{n=1}^{24}\xi^n$$
$$- \sum_{m,n=1}^{24} H_{nm}\xi^n\xi^m \tag{2.47}$$

where $\mathbb{S}$, σ, T, T_0, $\hat{e}^n$, and ξ^n are the compliance tensor, local stress tensor, temperature, reference temperature, transformation strain, and martensitic volume fraction on nth variant, respectively. The parameter β is a material constant. The last term in (2.47) is an approximation which accounts for martensite variant interactions by introducing an interaction matrix H [55]. The interaction matrix H indicates the variants with high and low interaction energies as given in Table 1 in Ref. [67].

By defining a driving force f_n, for each variant as

$$f_n = \frac{\partial G}{\partial \xi_n} = \sigma : \hat{e}^n - \sum_{m=1}^{24} H_{nm}\xi^m - \beta(T - T_0) \tag{2.48}$$

the criteria for forward transformation of austenite to nth martensite variant is given by $f_n = f^{am}$, and the condition for reverse transformation of nth martensite variant to austenite is expressed as $f_n = f^{ma}$, where f^{am} and f^{ma} are critical energies for A to M and M to A transformations, respectively. During the forward and reverse phase transformations, the consistency condition $\dot{f}_n = 0$ is used for obtaining the flow rule for the martensitic transformation as

$$\dot{\sigma} : \hat{e}^n = \sum_{m \in \mathbf{Q}} H_{nm}\dot{\xi}^m + \beta\dot{T} \tag{2.49}$$

where $\mathbf{Q} = \{Q_1, ..., Q_q\}$, $1 \leq q \leq 24$ is a set of q numbers corresponding the active variants that satisfy the forward or reverse transformation conditions mentioned for the deriving force (2.48).

2.2.2 Finite Element Modeling of Polycrystalline SMAs

The introduced micromechanical framework can be implemented into a numerical mode, such as finite element, to study polycrystalline SMAs. For developing an incremental displacement-based finite element, in addition to the constitutive relations given in the previous section, the tangent stiffness (mechanical Jacobian) and thermal moduli tensors are also needed. In order to obtain these tensors, the constitutive model of SMA material should be linearized and represented as an incremental form. Details of deriving the mechanical and temperature Jacobians are expressed in Ref. [67]. We will improve the mentioned micromechanical model by considering a more realistic thermomechanical coupling in deriving the Jacobians in the sequel.

For an infinitesimal time increment, Δt, the time rate of each parameter $\mathcal{P}$ can be approximated by $\dot{\mathcal{P}} = \Delta \mathcal{P} / \Delta t$. Using this approximation, (2.49) can be written as a set of q simultaneous equations ($1 \le q \le 24$ is the number of variants that satisfy the transformation conditions (2.48))

$$\Delta \sigma : \hat{e}^n = \sum_{m=1}^{24} H_{nm} \Delta \xi^m + \beta \Delta T \tag{2.50}$$

Using this approximation, the stress increment is related linearly to the increment of temperature and the martensitic volume fraction of active variants. It can be shown that the martensitic volume fraction increment is also related to the strain increment linearly. The stress increment can be approximated by

$$\Delta \sigma = \frac{\partial \Delta \sigma}{\partial \Delta \epsilon} : \Delta \epsilon + \frac{\partial \Delta \sigma}{\partial \Delta T} \Delta T \tag{2.51}$$

where $\partial \Delta \sigma / \partial \Delta \epsilon$ is the mechanical Jacobian. The stress increment can be written as

$$\Delta \sigma = \mathbb{S}^{-1} : \left(\Delta \epsilon - \sum_{m=1}^{24} \Delta \epsilon^m \right) \tag{2.52}$$

where ϵ^i is the contribution of transformation strain from the ith martensite variant that is related to the stress-free transformation strain of the corresponding variant through the volume fraction as $\epsilon^i = \hat{\epsilon}^i \xi^i$. Substituting (2.52) into (2.50), one obtains

$$\mathbb{S}^{-1} : \left(\Delta \epsilon - \sum_{m=1}^{24} \hat{\epsilon}^m \Delta \xi^m \right) : \hat{e}^n = \sum_{m=1}^{24} H_{nm} \Delta \xi^m + \beta \Delta T \tag{2.53}$$

By defining a transformation matrix Γ and a driving force vector F as

$$\Gamma_{mn} = \hat{\epsilon}^m : \mathbb{S}^{-1} : \hat{\epsilon}^n + H_{mn}, \quad F_m = \hat{\epsilon}^m : \mathbb{S}^{-1} : \Delta\epsilon - \beta\Delta T \tag{2.54}$$

the set of q simultaneous equations (2.53) can be rewritten in the matrix form as

$$[\Gamma]\{\Delta\xi\} = \{F\} \tag{2.55}$$

where the size of vectors and the transformation matrix depends on the number of active variants (q) and $\{\Delta\xi\}$ is a vector containing the incremental change of volume fraction of all the active variants. This set of equations will be used for calculating the incremental change of volume fractions.

For deriving the mechanical Jacobian, the incremental stress–strain relation (2.52) is differentiated with respect to the strain increment as

$$\frac{\partial\Delta\sigma}{\partial\Delta\epsilon} = \mathbb{S}^{-1} : \left(\mathbb{I} - \sum_{m=1}^{24} \frac{\partial\Delta\epsilon^m}{\partial\Delta\epsilon} \right) \tag{2.56}$$

where $\mathbb{I}$ is the identity tensor. The derivative of the transformation strain with respect to the total strain can be written as

$$\frac{\partial\Delta\epsilon^m}{\partial\Delta\epsilon} = \frac{\partial\Delta\epsilon^m}{\partial\Delta\xi^m} \otimes \frac{\partial\Delta\xi^m}{\partial\Delta\epsilon} = \epsilon^m \otimes \Gamma_{mn}^{-1} \frac{\partial F^n}{\partial\Delta\epsilon} \tag{2.57}$$

where $\otimes$ denotes the tensor product, and the last term is derived using the inverse of (2.55). Substituting (2.57) into (2.56) gives the mechanical Jacobian to be implemented in the finite element formulation. The thermal Jacobian is obtained by differentiating (2.52) with respect to the temperature increment as

$$\frac{\partial\Delta\sigma}{\partial\Delta T} = \mathbb{S}^{-1} : \left(-\sum_{m=1}^{24} \frac{\partial\Delta\epsilon^m}{\partial\Delta T} \right) \tag{2.58}$$

where the derivative of the transformation strain with respect to temperature is calculated using the chain rule as

$$\frac{\partial\Delta\epsilon^m}{\partial\Delta T} = \frac{\partial\Delta\epsilon^m}{\partial\Delta\xi^m} \frac{\partial\Delta\xi^m}{\partial\Delta T} = \hat{\epsilon}^m \Gamma_{mn}^{-1} \frac{\partial F^n}{\partial\Delta T} \tag{2.59}$$

Substituting (2.59) into (2.58) gives the thermal Jacobian.

2.2.3 Thermomechanical Coupling in the Micromechanical Model

A comprehensive description of the energy balance equation can be used for obtaining the thermal coupled equations at microscale. We use a similar method as used in the previous sections for obtaining the governing thermo-mechanical equations based on phenomenological constitutive models. The coupled thermomechanical governing equations at microscale are derived by considering the first law of thermodynamics in local form (2.38), the dissipation inequality (2.39), and the second law of thermodynamics (2.40). Using the definition of the Gibbs free energy (2.47), the time derivative of G is given by

$$\dot{G} = \frac{\partial G}{\partial \boldsymbol{\sigma}} : \dot{\boldsymbol{\sigma}} + \frac{\partial G}{\partial T}\dot{T} + \sum_{n=1}^{24} \frac{\partial G}{\partial \xi^n} : \dot{\xi}^n \qquad (2.60)$$

Substituting (2.60) into (2.40) gives

$$-\left(\frac{\partial G}{\partial \boldsymbol{\sigma}} + \epsilon\right) : \dot{\boldsymbol{\sigma}} - \left(\frac{\partial G}{\partial T} + \rho s\right)\dot{T} - \sum_{n=1}^{24} \frac{\partial G}{\partial \xi^n} : \dot{\xi}^n \geq 0 \qquad (2.61)$$

Validity of (2.61) for all $\dot{\boldsymbol{\sigma}}$ and $\dot{T}$ implies the following constitutive equations

$$-\frac{\partial G}{\partial \boldsymbol{\sigma}} = \epsilon, \quad -\frac{\partial G}{\partial T} = \rho s \qquad (2.62)$$

The energy balance equation is obtained by substituting (2.62) and (2.60) into (2.38) as

$$\rho T \dot{s} = -\sum_{n=1}^{24} \frac{\partial G}{\partial \xi^n} \dot{\xi}^n - \operatorname{div} q + \rho \hat{g} \qquad (2.63)$$

The constitutive relation $(2.62)_2$ is used for calculating the time derivative of the specific entropy as

$$\rho \dot{s} = -\frac{\partial \dot{G}}{\partial T} = -\frac{\partial^2 G}{\partial \boldsymbol{\sigma} \partial T} : \dot{\boldsymbol{\sigma}} - \frac{\partial^2 G}{\partial T^2}\dot{T} - \sum_{n=1}^{24} \frac{\partial^2 G}{\partial \xi^n \partial T} \dot{\xi}^n \qquad (2.64)$$

which after substituting (2.47) into (2.64) gives the rate of change of specific entropy as

$$\rho \dot{s} = \alpha : \dot{\sigma} + \frac{\rho c}{T}\dot{T} - \beta \sum_{n=1}^{24} \dot{\xi}^n \qquad (2.65)$$

Substituting (2.65) into (2.63), the final form of the first law is obtained as

$$\rho c \dot{T} = \sum_{n=1}^{24} \left(-\frac{\partial G}{\partial \xi^n} + \beta T \right) \dot{\xi}^n - T\alpha : \dot{\sigma} - \operatorname{div} q + \rho \hat{g} \qquad (2.66)$$

The energy balance (2.66) is used for finding the volumetric heat generation in SMAs as

$$\mathcal{R} = \sum_{n=1}^{24} \left(-\frac{\partial G}{\partial \xi^n} + \beta T \right) \dot{\xi}^n - T\alpha : \dot{\sigma} = \sum_{n=1}^{24} (f_n + \beta T)\dot{\xi}^n - T\alpha : \dot{\sigma} \qquad (2.67)$$

where the term f_n is defined in (2.48). It is worth noting that during the phase transformation, the consistency condition implies that $f_n = f^\pm$, where $f^+ = f^{am}$ for the forward phase transformation and $f^- = f^{ma}$ for the reverse transformation. The constitutive equations, Jacobians, volumetric heat generation $\mathcal{R}$, and its derivatives with respect to the temperature and strain are given to the finite element model in a user subroutine (UMAT) for developing a coupled thermomechanical model. This subroutine is written in a local coordinate system. In the finite element model, a separate local orientation can be assigned to the elements in each grain. All the orientation-dependent parameters are transformed to the local coordinate before passing them into the subroutine and transformed back to the global coordinate when the subroutine results are given to the finite element code.

In some of the previously reported works in the literature that study polycrystalline SMAs using a micromechanical-based model, the polycrystal structure was designed for resembling the microstructure as closely as possible to the actual structure (see Figure 2 in Ref. [67] and Figure 5 in Ref. [68]). In these works, hexagonal prisms and cubes are used for modeling the geometry of each grain which is far from the actual grain shape. However, a more realistic finite element mesh for modeling the polycrystal structure in SMAs can be developed using Voronoi tessellations that are widely accepted to model polycrystalline aggregates, as they provide a realistic approximation of the actual

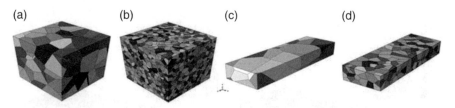

Figure   2.7 Three-dimensional   Voronoi   tessellations   for   modeling   a polycrystalline SMAs with (a) 100, (b) 5000, (c) 26, and (d) 357 grains

microstructure of nonuniform grain shapes [69–72]. Some examples of constructed polycrystalline models with different number of grains are shown in Figure 2.7. Models (a) and (b) with 100 and 5000 grains are used for studying simple tension. Models (c) and (d) with 26 and 537 grains are created for studying the bending of polycrystalline SMA microbeams.

This model can be used for studying the role of texture (defined as the distribution on crystal orientations in all the grains in the polycrystalline sample), grain size, and loading rate on the response of polycrystalline SMAs subjected to uniaxial loading or bending besides studying microscaled SMA devices with various shapes subjected to different loading conditions. As an example, consider bending of a microscaled polycrystalline SMA beam. Such microbeams can be produced by FIB micromachining of bulk material. Several works have been reported on studying SMA micropillars subjected to various loading conditions and in situ scanning with SEM during the loading to study the response of the material at microscale [7–12]. The presented model in this section can be used to study several important aspects of the thermomechanical response of SMAs at microscale, some of which are practically very difficult to be studied in the experiments. For example, microscaled beams (with 357 grains as shown in Figure 2.7c and d) with different distributions of crystal orientations are studied as shown in Figure 2.8. The crystal orientations in the grains are randomly distributed in the untextured sample, while the [111] directions of the crystal lattice among all the grains are dominantly parallel to the beam axis in the textured sample.

The beam is clamped at one edge, and the other end is deformed in transverse direction. The martensitic volume fraction (summation of volume fraction of all active martensite variants) is shown in Figure 2.8a and b for the textured and untextured beams, respectively. Comparing the volume fraction distribution near the clamped edge in these figures shows the asymmetric distribution of martensite volume fraction in the textured beam. It has been previously shown that while an untextured NiTi polycrystalline response

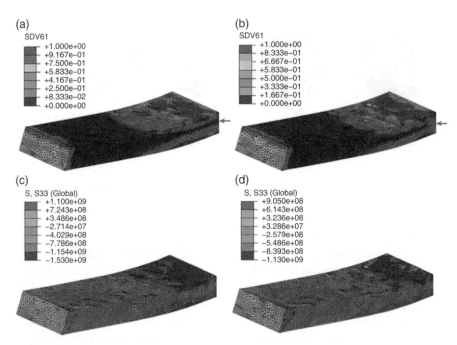

Figure 2.8 (a and b) Martensitic volume fraction distribution and (c and d) normal stress distributions in a polycrystal beam with 357 grains subjected to bending with (a and c) textured and (b and d) untextured crystal orientations

is semisymmetric in tension–compression, the tensile and compressive responses are remarkably different in a textured polycrystal [1]. This happens because in the textured material, the majority of grains are oriented along the [111] crystallographic direction, which is soft under tensile loading and hard under compression. In the untextured samples, the crystallographic directions are randomly distributed and that leads to an almost symmetric response in tension–compression. As shown in Figure 2.8a, the neutral axis position (marked with an arrow on the figure) is shifted toward the compressive part of the cross section for the textured beam while it is near the center line for the untextured beam due to the symmetry in tension–compression response. Comparing the results of this section with those presented in the previous sections based on the macroscopic models shows that the J_2-based phenomenological models are suitable for modeling the response of untextured polycrystalline materials and J_2–I_1-based models can be used for studying the textured polycrystal SMAs in bending with an acceptable accuracy.

The normal stress distribution is compared for the textured and untextured beams in Figure 2.8c and d. An asymmetry in the stress distribution for the textured material is observed, and it is shown that the maximum compressive stress is remarkably larger than the maximum tensile stress in the textured beam.

2.3 Summary

This chapter presented all major methods for modeling the behavior of shape memory alloys. Phenomenological approach of modeling received more attention as a practical tool for engineering application. Different phenomenological models were discussed, and the methodology for finding the parameters of these models was presented. The standards for performing the experiments for parameter identification were also introduced. Simulation techniques for numerical evaluation of the behavior of the material were introduced followed by simulation methods, which are specifically suitable for SMA actuation. In these actuators, the behavior of the alloys is affected by the dynamics as well as kinematics of the system. Particular attention was paid to the investigation of the coupled dynamics between the actuator and system.

References

[1] Gall K, Sehitoglu H. The role of texture in tension-compression asymmetry in polycrystalline NiTi. International Journal of Plasticity. 1999;**15**(1):69–92.

[2] Anand L, Gurtin ME. Thermal effects in the superelasticity of crystalline shape-memory materials. Journal of the Mechanics and Physics of Solids. 2003;**51**(6): 1015–1058.

[3] Kahn H, Huff MA, Heuer AH. The TiNi shape-memory alloy and its applications for MEMS. Journal of Micromechanics and Microengineering. 1998;**8**(3):213–221.

[4] Bhattacharya K, James R. The material is the machine. Science. 2005;**307**(5706): 53–54.

[5] Shin DD, Mohanchandra KP, Carman GP. Development of hydraulic linear actuator using thin film SMA. Sensors and Actuators, A: Physical. 2005;**119**(1): 151–156.

[6] Krulevitch P, Lee AP, Ramsey PB, Trevino JC, Hamilton J, Northrup MA. Thin film shape memory alloy microactuators. Journal of Microelectromechanical Systems. 1996;**5**(4):270–282.

[7] Volkert CA, Minor AM. Focused ion beam microscopy and micromachining. MRS Bulletin. 2007;**32**(5):389–395.

[8] Frick CP, Orso S, Arzt E. Loss of pseudoelasticity in nickel-titanium sub-micron compression pillars. Acta Materialia. 2007;**55**(11):3845–3855.

[9] San Juan JM, No ML, Schuh CA. Superelasticity and shape memory in micro- and nanometer-scale pillars. Advanced Materials. 2008;**20**(2):272–278.

[10] Manjeri RM, Qiu S, Mara N, Misra A, Vaidyanathan R. Superelastic response of [111] and [101] oriented NiTi micropillars. Journal of Applied Physics. 2010; **108**(2):023501.

[11] Juan JS, No ML, Schuh CA. Nanoscale shape-memory alloys for ultrahigh mechanical damping. Nature Nanotechnology. 2009;**4**(7):415–419.

[12] Clark BG, Gianola DS, Kraft O, Frick CP. Size independent shape memory behavior of nickel-titanium. Advanced Engineering Materials. 2010;**12**(8):808–815.

[13] Kim JY, Jang D, Greer JR. Insight into the deformation behavior of niobium single crystals under uniaxial compression and tension at the nanoscale. Scripta Materialia. 2009;**61**(3):300–303.

[14] Kim JY, Greer JR. Tensile and compressive behavior of gold and molybdenum single crystals at the nano-scale. Acta Materialia. 2009;**57**(17):5245–5253.

[15] Sato T, Saitoh KI, Shinke N. Molecular dynamics study on microscopic mechanism for phase transformation of Ni-Ti alloy. Modelling and Simulation in Materials Science and Engineering. 2006;**14**(5):S39–S46.

[16] Saitoh Ki, Sato T, Shinke N. Atomic dynamics and energetics of martensitic transformation in nickel-titanium shape memory alloy. Materials transactions. 2006; **47**(3):742–749.

[17] Mutter D, Nielaba P. Simulation of structural phase transitions in NiTi. Physical Review B. 2010;**82**(22):224201.

[18] Mutter D, Nielaba P. Simulation of the shape memory effect in a NiTi nano model system. Journal of Alloys and Compounds. 2013;**577**, Suppl. 1:S83–S87.

[19] Zhong Y, Gall K, Zhu T. Atomistic study of nanotwins in NiTi shape memory alloys. Journal of Applied Physics. 2011;**110**(3):6301–6311.

[20] Zhong Y, Gall K, Zhu T. Atomistic characterization of pseudoelasticity and shape memory in NiTi nanopillars. Acta Materialia. 2013;**60**:6301–6311.

[21] Kondo Y, Ru Q, Takayanagi K. Thickness induced structural phase transition of gold nanofilm. Physical Review Letters. 1999;**82**(4):751–754.

[22] Kondo Y, Takayanagi K. Synthesis and characterization of helical multi-shell gold nanowires. Science. 2000;**289**(5479):606–608.

[23] Hasmy A, Medina E. Thickness induced structural transition in suspended fcc metal nanofilms. Physical Review Letters. 2002;**88**(9):961031–961034.

[24] Diao JK, Gall K, Dunn ML. Surface-stress-induced phase transformation in metal nanowires. Nature Materials. 2003;**2**(10):656–660.

[25] Mirzaeifar R, Gall K, Zhu T, Yavari A, DesRoches R. Structural transformations in NiTi shape memory alloy nanowires. Journal of Applied Physics. 2014;**115**(19): 194307.

[26] Artemev A, Wang Y, Khachaturyan AG. Three-dimensional phase field model and simulation of martensitic transformation in multilayer systems under applied stresses. Acta Materialia. 2000;**48**(10):2503–2518.

[27] Falk F. Model free energy, mechanics and thermodynamics of shape memory alloys. Acta Metallurgica. 1980;28(12):1773–1780.

[28] Ichitsubo T, Tanaka K, Koiwa M, Yamazaki Y. Kinetics of cubic to tetragonal transformation under external field by the time-dependent Ginzburg-Landau approach. Physical Review B. 2000;62:5435–5441.

[29] Rasmussen KO, Lookman T, Saxena A, Bishop AR, Albers RC, Shenoy SR. Three-dimensional elastic compatibility and varieties of twins in martensites. Physical Review Letters 2001;87:055704.

[30] Levitas VI, Preston DL. Three-dimensional Landau theory for multivariant stress-induced martensitic phase transformations.I. Austenite-martensite. Physical Review B. 2002;66:134206.

[31] Levitas VI, Preston DL. Three-dimensional Landau theory for multivariant stress-induced martensitic phase transformations.II. Multivariant phase transformations and stress space analysis. Physical Review B. 2002;66:134207.

[32] Levitas VI, Preston DL, Lee DW. Three-dimensional Landau theory for multivariant stress-induced martensitic phase transformations.III. Alternative potentials, critical nuclei, kink solutions, and dislocation theory. Physical Review B. 2002;66:134201.

[33] Hormann K, Zimmer J. On Landau theory and symmetric energy landscapes for phase transitions. Journal of the Mechanics and Physics of Solids. 2007;55(7): 1385–1409.

[34] Birman V. Review of mechanics of shape memory alloy structures. Applied Mechanics Reviews. 1997;50(1):629–645.

[35] Lagoudas DC, Entchev PB, Popov P, Patoor E, Brinson LC, Gao X. Shape memory alloys, Part II: Modeling of polycrystals. Mechanics of Materials. 2006;38(5–6): 430–462.

[36] Lagoudas DC. Shape memory alloys: modeling and engineering applications. Springer: New York; 2008.

[37] Liang C, Rogers CA. The multi-dimensional constitutive relations of shape memory alloys. Journal of Engineering Mathematics. 1992;26:429–443.

[38] Tanaka K, Nishimura F, Hayashi T, Tobushi H, Lexcellent C. Phenomenological analysis on subloops and cyclic behavior in shape memory alloys under mechanical and/or thermal loads. Mechanics of Materials. 1995;19:281–292.

[39] Boyd JG, Lagoudas DC. Thermodynamical constitutive model for shape memory materials. Part I. The monolithic shape memory alloy. International Journal of Plasticity. 1996;12(6):805–842.

[40] Lagoudas DC, Bo Z, Qidwai MA. Unified thermodynamic constitutive model for SMA and finite element analysis of active metal matrix composites. Mechanics of composite materials and structures. 1996;3(2):153–179.

[41] Qidwai MA, Lagoudas DC. Numerical implementation of a shape memory alloy thermomechanical constitutive model using return mapping algorithms. International Journal for Numerical Methods in Engineering. 2000;47(6):1123–1168.

[42] Auricchio F, Sacco E. A temperature-dependent beam for shape-memory alloys: constitutive modelling, finite-element implementation and numerical

simulations. Computer Methods in Applied Mechanics and Engineering. 1999;**174**:171–190.

[43] Marfia S, Sacco E, Reddy JN. Superelastic and shape memory effects in laminated shape-memory-alloy beams. AIAA Journal. 2003;**41**(1):100–109.

[44] Qidwai MA, Lagoudas DC. On thermomechanics and transformation surfaces of polycrystalline NiTi shape memory alloy material. International journal of plasticity. 2000;**16**(10):1309–1343.

[45] Mirzaeifar R, Desroches R, Yavari A. Analysis of the rate-dependent coupled thermo-mechanical response of shape memory alloy bars and wires in tension. Continuum Mechanics and Thermodynamics. 2011;**23**(4):363–385.

[46] Bo Z, Lagoudas DC. Thermomechanical modeling of polycrystalline SMAs under cyclic loading, Part I: Theoretical derivations. International Journal of Engineering Science. 1999;**37**(9):1089–1140.

[47] Cardano G, Witmer TR, Ore O. The Rules of Algebra: (ars Magna). Dover Publications. 2007.

[48] Mirzaeifar R, Desroches R, Yavari A. Exact solutions for pure torsion of shape memory alloy circular bars. Mechanics of Materials. 2010;**42**(8):797–806.

[49] Wahl AM. Mechanical springs. Penton publishing company: Cleveland; 1944.

[50] Gillet Y, Patoor E, Berveiller M. Calculation of pseudoelastic elements using a non-symmetrical thermomechanical transformation criterion and associated rule. Journal of Intelligent Material Systems and Structures. 1998;**9**:366–378.

[51] Auricchio F, Taylor RL, Lubliner J. Shape-memory alloys: macromodelling and numerical simulations of the superelastic behavior. Computer Methods in Applied Mechanics and Engineering. 1997;**146**:281–312.

[52] Liu Y, Xie Z, Humbeeck JV, Delaey L. Asymmetry of stress-strain curves under tension and compression for NiTi shape memory alloys. Acta Materialia. 1998;**46**(12):4325–4338.

[53] Gall K, Sehitoglu H, Chumlyakov YI, Kireeva IV. Tension-compression asymmetry of the stress-strain response in aged single crystal and polycrystalline NiTi. Acta Materialia. 1999;**47**(4):1203–1217.

[54] Gall K, Sehitoglu H, Anderson R, Karaman I, Chumlyakov YI, Kireeva IV. On the mechanical behavior of single crystal NiTi shape memory alloys and related polycrystalline phenomenon. Materials Science and Engineering A. 2001;**317**(1–2):85–92.

[55] Patoor E, Elamrani M, Eberhardt A, Berveiller M. Determination of the origin for the dissymmetry observed between tensile and compression tests on shape-memory alloys. Journal De Physique IV. 1995;**5**(C2):495–500.

[56] Paiva A, Savi MA, Braga AMB, Pacheco PMCL. A constitutive model for shape memory alloys considering tensile-compressive asymmetry and plasticity. International Journal of Solids and Structures. 2005;**42**(11–12):3439–3457.

[57] Auricchio F, Reali A, Stefanelli U. A macroscopic 1D model for shape memory alloys including asymmetric behaviors and transformation-dependent elastic properties. Computer Methods in Applied Mechanics and Engineering. 2009;**198**(17–20):1631–1637.

[58] Mirzaeifar R, Desroches R, Yavari A, Gall K. On superelastic bending of shape memory alloy beams. International Journal of Solids and Structures. 2013; 50(10):1664–1680.

[59] Abramowitz M, Stegun IA. Handbook of mathematical functions with formulas, graphs and mathematical tables. National Bureau of Standards Applied Mathematics: Washington; 1964.

[60] Jacobus K, Sehitoglu H, Balzer M. Effect of stress state on the stress-induced martensitic transformation in polycrystalline Ni-Ti alloy. Metallurgical and Materials Transactions A. 1996;27:3066–3073.

[61] Airoldi G, Riva G, Rivolta B, Vanelli M. DSC calibration in the study of shape memory alloys. Journal of Thermal Analysis and Calorimetry. 1994;42:781–791.

[62] He XM, Rong LJ. DSC analysis of reverse martensitic transformation in deformed TiNiNb shape memory alloy. Scripta Materialia. 2004;51(1):7–11.

[63] Desroches R, McCormick J, Delemont M. Cyclic properties of superelastic shape memory alloy wires and bars. Journal of Structural Engineering. 2004; 130(1):38–46.

[64] McCormick J, Tyber J, DesRoches R, Gall K, Maier HJ. Structural engineering with NiTi. II: Mechanical behavior and scaling. Journal of Engineering Mechanics. 2007;133(9):1019–1029.

[65] Mirzaeifar R, Desroches R, Yavari A, Gall K. Coupled thermo-mechanical analysis of shape memory alloy circular bars in pure torsion. International Journal of Non-Linear Mechanics. 2012;47(3):118–128.

[66] Andani MT, Alipour A, Elahinia M. Coupled rate-dependent superelastic behavior of shape memory alloy bars induced by combined axial-torsional loading: a semi-analytic modeling. Journal of Intelligent Material Systems and Structures. 2013;24(16):1995–2007.

[67] Lim TJ, McDowell DL. Cyclic thermomechanical behavior of a polycrystalline pseudoelastic shape memory alloy. Journal of the Mechanics and Physics of Solids. 2002;50(3):651–676.

[68] Thamburaja P, Anand L. Polycrystalline shape-memory materials: effect of crystallographic texture. Journal of the Mechanics and Physics of Solids. 2001; 49(4):709–737.

[69] Boots BN. The arrangement of cells in random networks. Metallography. 1982; 15(1):53–62.

[70] Fritzen F, Böhlke T, Schnack E. Periodic three-dimensional mesh generation for crystalline aggregates based on Voronoi tessellations. Computational Mechanics. 2009;43:701–713.

[71] Zhang P, Balint D, Lin J. An integrated scheme for crystal plasticity analysis: virtual grain structure generation. Computational Materials Science. 2011; 50(10):2854–2864.

[72] Simonovski I, Cizelj L. Automatic parallel generation of finite element meshes for complex spatial structures. Computational Materials Science. 2011;50(5): 1606–1618.

3

SMA Actuation Mechanisms

Masood Taheri Andani, Francesco Bucchi and Mohammad H. Elahinia

This chapter builds on the mathematical foundation developed in previous chapters to discuss design methodologies for several SMA actuators. The basic step in design is defining the requirements for linear or angular displacement and the force or torque that the actuator should deliver. These requirements in conjunction with the unique thermomechanical properties of the alloys are used to design the actuation mechanism for the system. The performance of the actuation can be evaluated based on the mathematical model of the system using simulation tools developed in Chapters 1 and 2. Unless special thermo-mechanical heat treatments are used, SMA devices produce one-way actuation. To create repetitive actuation, therefore, a bias mechanism is necessary to restore the geometry of the actuator element to its detwinned martensite form. To this end, different actuation mechanisms have been developed using various bias methods to achieve repeatable motion. Some of these mechanisms include the use of energy storing springs, other SM elements, and super-elastic elements. While bias spring actuators are simpler to fabricate, the control of the resulting actuator is more challenging, as the actuation remains one directional. The other downside of this is that a significant portion of the SMA actuation force is consumed to overcome the linearly increasing force of

Shape Memory Alloy Actuators: Design, Fabrication, and Experimental Evaluation, First Edition. Mohammad H. Elahinia.
© 2016 John Wiley & Sons, Ltd. Published 2016 by John Wiley & Sons, Ltd.

(a)   (b)

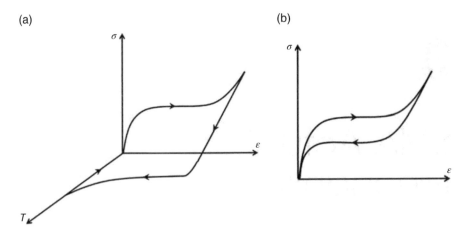

Figure 3.1 Mechanical response of a (a) shape memory and a (b) superelastic antagonist element includes a plateau that, in an antagonistic arrangement, reduces the resistance to actuation of the active element

the spring. This latter limitation is resolved by the use of shape memory (SM) or superelastic antagonistic elements. The resistance force of such elements, due to reorientation or phase transformation, follows a hysteretic loop with a close-to-zero stiffness during the plateau as shown in Figure 3.1.

3.1 Bias-Type Actuators

In this section, design and analysis of SMA actuators for two rotary systems are discussed. In the vast majority of SMA actuators, the SM element undergoes simple tension. A one-degree of-freedom rotary SMA actuator that was introduced in Chapter 1 and is shown in Figure 3.2 is an example. In this device, motion is generated through Joule heating of an SMA wire which is guided through the pulleys. The reverse motion takes place when the wire is cooled and the amount of tensile stress applied by the bias spring and the weight of the payload is large enough to elongate the wire. The length of the SMA wire, based on Equation 1.1, is selected so that upon a change of strain of −4%, the arm rotates from its initial position at −45° to the top position at +90°. The diameter of the wire is selected, based on Equation 1.2, such that at the maximum allowable stress, the force applied by the SMA wire generates a large enough moment to overcome those of spring and payload while accelerating the moving arm.

The SMA-actuated system is modeled and simulated considering the phase transformation kinetics, heat transfer, constitutive model, dynamics,

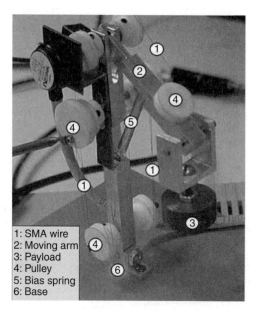

1: SMA wire
2: Moving arm
3: Payload
4: Pulley
5: Bias spring
6: Base

Figure 3.2 A one-degree-of-freedom rotary SMA actuator; the upward motion is generated by shape memory effect and the downward motion by the bias spring and the weight of the payload. From Refs. [1–5]

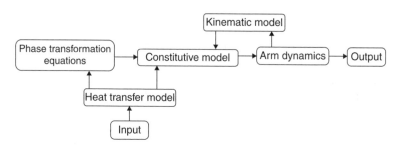

Figure 3.3 Block diagram of the sections of the lumped parameter model of the 1 DOF SMA actuator follows the causality of the variables. From Ref. [3]

kinematics, and electrical behavior. The main input of the model is the electrical voltage, and the output of the system is the angular rotation. The model also calculates the temperature, martensite fraction, strain, stress, and angular velocity. The submodels are bilaterally connected, forming an algebraic

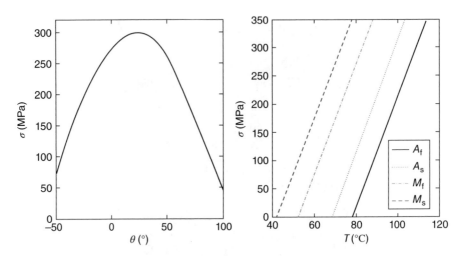

Figure 3.4 The stress of SMS wire changes as the arm rotates, causing a similar change in the four transformation temperatures

loop (see Figure 3.3). One reason for this interdependency is that physical properties of the SMA wire are functions of martensite fraction, stress, and temperature. More specifically, bilateral causality exists between dynamic and constitutive equations and also between constitutive and phase transformation equations, as was shown in Chapter 1.

What distinguishes this rotary system from the other SMA actuators is the fact that during the actuation, due to the kinematics of the system, the stress of the wire changes. As shown in Figure 3.4, this phenomenon leads to variation in the four transformation temperatures that together define the actuation limits. The changes in the four transformation temperatures are $A'_s = A_s + \frac{\delta\sigma}{C_A}$, $A'_f = A_f + \frac{\delta\sigma}{C_A}$, $M'_s = M_s + \frac{\delta\sigma}{C_M}$, and $M'_f = M_f + \frac{\delta\sigma}{C_M}$. It is therefore important for the model to capture the possible complex thermomechanical loadings that arise in such systems. Three of such possibilities are shown in Figure 3.5. In all these cases, the transformation temperatures change as the arm rotates over its range of motion. Even though the final temperature of the wire is different in these three cases, it is expected that the wire will completely transform to austenite phase and the arm will rotate to the top position. The relative value of these temperatures and the temperature of the SMA element define the state of material as was highlighted in the phase transformation kinetics in Chapter 1.

Figure 3.6 shows the transformation behavior of the SMA wire on the phase diagram. Starting from point A (martensite, arm at −45°), the temperature

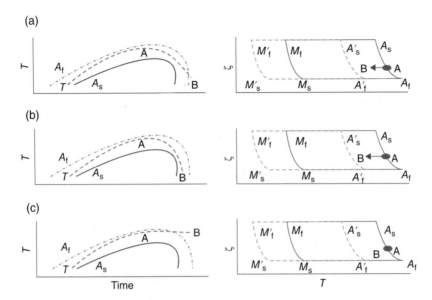

Figure 3.5 In the rotary SMA actuator, due to variation of stress during actuation, the four transformation temperatures change. Three possibilities are shown when heating the SMA element: (a) exceeding the austenite final temperature while cooling, (b) remaining in the phase transformation range while cooling, and (c) exceeding austenite final temperature with constant temperature

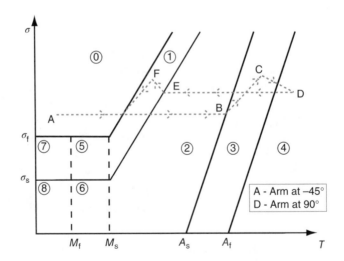

Figure 3.6 Phase diagram of the one-degree-of-freedom rotary bias-type actuator

increases, and at point B, the martensite to austenite transformation starts. The arm rises and the stress increases up to point C and then decreases as the arm passes through the maximum stress position and the arm approaches the vertical position. When the temperature decreases, a similar behavior occurs: at point E, the martensite transformation starts, and the stress increases to the maximum stress value defined by the kinematics and dynamics of the arm (point F) and then decreases as the arm approaches the initial position (−45°) and initial length of the wire is restored.

3.2 Antagonistic SM–SM Actuation

Two antagonistic SMA actuators can provide more precise position control. An example of such a system is shown Figure 3.7 where two pairs of SMA wires rotate a mirror about each axis of rotation [6, 7]. The amount of force is proportional to the diameter of the wire. To increase the fatigue life of the wire, as shown in Chapter 5, it is essential to limit the strain of the SMA actuator to 4%. This limitation, along with the required displacement of the actuator, defines the length of the wire. When in the middle position and before actuation, each of the antagonistic wires consists of an equal amount of temperature-induced and stress-induced martensite. When a wire is heated above the austenite final, it recovers the strain, becomes shorter, and transforms to the austenite phase. This causes rotation of the mirror about the corresponding axis. Simultaneously, the opposing wire is detwinned and transforms to the stress-induced martensite phase. At the end of actuation and when the active

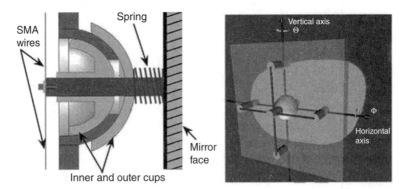

Figure 3.7 Each pair of SMA wires (left) forms an antagonistic actuator for rotating a mirror (right) about one of its axis of rotation; these four wires rotate about two axes. From Refs. [6, 7]

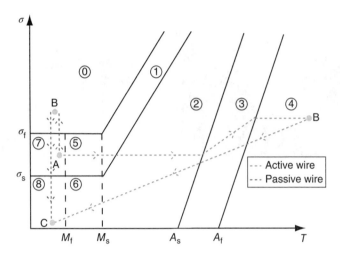

Figure 3.8 Phase diagram of the two antagonistic SMA wires in the mirror actuator

wire cools down, it transforms to the temperature-induced martensite phase, while maintaining the short length. This wire directly transforms to stress-induced martensite and the long length when the opposing wire is actuated. Because each of the antagonistic elements can be activated separately, a more precise position control is possible. In this actuator, to improve the bandwidth of the actuation, it is possible to keep a passive wire heated below the actuation start temperature. The additional benefit of using the opposing SMA elements, as shown in Figure 3.1, is that during detwinning the bias force is not proportionally increased and therefore does not reduce the actuation force. This leads to a higher efficiency in the actuation as a larger percentage of the force of the actuating wire is used for actuation.

Figure 3.8 shows the phase diagram of the transformation for both wires. It is worth noting that starting from the same phase state (point A) of stress-induced and temperature-induced martensite for both wires, the final state of the two wires is different (point C): the active wire is in temperature-induced martensite, whereas the passive wire is in stress-induced martensite.

3.3 SM Spring Actuation

SMA spring actuators can be fabricated by shape setting wires. As explained in Chapter 6, in this process, the wire is constrained on a mandrel before the

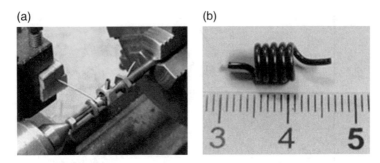

Figure 3.9 (a) Wrapping procedure using a lathe of a (b) nitinol spring (ruler in mm)

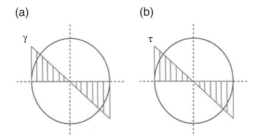

Figure 3.10 (a) Shear strain and (b) shear stress in the cross section of the spring wire of a linear material

heat treatment. Figure 3.9 shows the wrapping procedure by a lathe [8]. A spring actuator, when compared to its wire counterpart, provides a larger range of motion in a smaller package with a reduced actuation force. The helicoidal spring elongation δ is related to the shear strain γ at the external radius of the wire cross section:

$$\delta = \frac{\pi N D^2}{d}\gamma \tag{3.1}$$

where N is the number of spring coils, D is the coil diameter, and d is the wire diameter. Figure 3.10 shows the wire shear strain and shear stress in conventional material in the elastic region, where the stress–strain relationship is linear.

In SMA springs, the distribution of shear strain, caused by the deformation of the wire, in the wire cross section is still considered linear. The shear stress

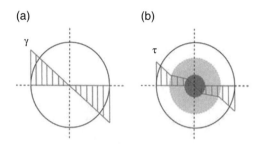

Figure 3.11 (a) Shear strain and (b) shear stress in the cross section of the spring wire of a SMA material

distribution on the cross section due to the nonlinearity of material behavior however is not linear. In order to understand the phase transformation within the wire, it is useful to introduce the equivalent stress (Cauchy) and the equivalent strain (Lagrangian) [9]:

$$\sigma_{eq} = \sqrt{\sigma^2 + 3\tau^2} \tag{3.2}$$

$$\varepsilon_{eq} = \sqrt{\varepsilon^2 + \frac{4}{3}\gamma^2} \tag{3.3}$$

where σ and ε are the tensile stress and strain, respectively, whereas τ and γ are the shear stress and strain, respectively. Using these equivalencies, the stress–strain–temperature can be generalized in terms of shear stress–shear strain temperature by replacing σ and ε with σ_{eq} and ε_{eq}.

Considering a spring made of NiTi and initially in the twinned martensite phase, if a sufficient load is applied, three different concentric regions can be identified within the wire cross section, as depicted in Figure 3.11.

In the internal area of the wire (gray), the stress is not enough to induce the detwinning of the martensite, so only twinned martensite is present ($\sigma_{eq} < \sigma_s^{cr}$, Figure 3.12), and the $\sigma_{eq}(r)$ varies linearly with the ε_{eq}. In the central area, both twinned and detwinned martensite are present ($\sigma_s^{cr} < \sigma_{eq} < \sigma_f^{cr}$), and the $\sigma_{eq}(r)$ is almost constant. Finally, in the external area ($\sigma_{eq} > \sigma_f^{cr}$), only detwinned martensite is present, and the $\sigma_{eq}(r)$ varies linearly with the ε_{eq}. Once the loading is completed and the force is removed, the deformation due to martensite detwinning remains, and it is recoverable by heating (inverse transformation). The maximum spring elongation δ_L due to martensite detwinning is

$$\delta_L = \frac{\pi N D^2}{d}\gamma_L = \frac{\pi N D^2}{d}\frac{\sqrt{3}}{2}\varepsilon_L \tag{3.4}$$

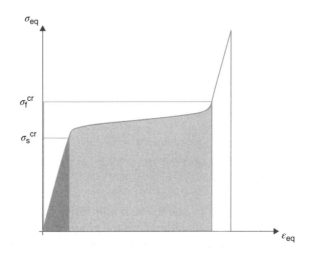

Figure 3.12 Equivalent stress versus equivalent strain of a NiTi wire during loading, starting from twinned martensite

where γ_L is the maximum shear residual strain, which can be related to the maximum residual strain ε_L by the Lagrangian equivalent strain definition. The term $\sqrt{3}/2$ appears because in the case of pure maximum residual shear strain ($\varepsilon = 0$, $\gamma = \gamma_L$), Equation 3.3 is

$$\varepsilon_L^{eq} = \sqrt{\frac{4}{3}\gamma_L^2} \qquad (3.5)$$

and if the maximum residual equivalent strain ε_L^{eq} is assumed to be equal to ε_L, γ_L is

$$\gamma_L = \frac{\sqrt{3}}{2}\varepsilon_L \qquad (3.6)$$

In order to assess the achievable deformation of a SMA spring, let's consider a spring made of NiTi with coils diameter $D = 5$ mm, wire diameter $d = 1$ mm, and $N = 5$ coils. If we consider the maximum residual strain $\varepsilon_L = 0.06$, the maximum spring elongation will be 20 mm, which is four times greater than the spring coil core (5 coils, 1 mm each one).

3.3.1 SM Spring-Actuated Device

In this section, as a representative example for SMA spring actuation, a hybrid SMA–magnetorheological (MR) fluid device is presented [10]. The device, shown in Figure 3.13, is composed of a permanent magnet MR clutch [11] coaxially built with a sliding spline sleeve, actuated by an antagonistic combination of SMA springs and bias springs. The MR clutch is composed of a primary shaft (no. 1) and a secondary shaft (no. 2), separated by a thin gap of MR fluid (no. 3). A permanent magnet (no. 4) can axially slide within this housing. The clutch has two states, determined by the magnet position: if the magnet is close to the fluid (right side of the housing, as represented in Figure 3.13), the clutch is engaged, due to the activation of MR fluid, and can transmit up to 3 N·m of torque. On the contrary, if the magnet is positioned to the left side of the housing, the MR fluid is not subject to a strong magnetic field and the clutch is disengaged. In this condition, the torque transmitted by the clutch is less than 0.5 N·m.

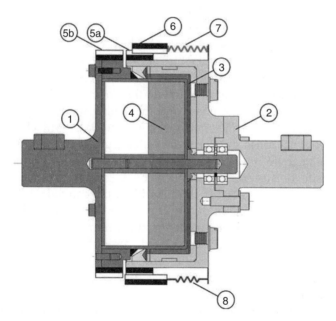

Figure 3.13 The hybrid SMA–MR clutch [10] has two modes of operation (low temperature and high temperature) to optimize the transferred torque

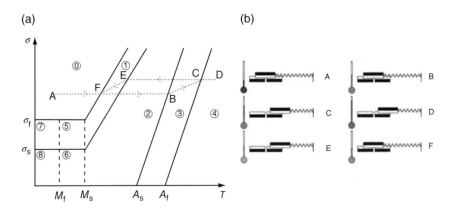

Figure 3.14 Phase diagram (a) and sleeve position (b) for the sliding spline sleeve actuated by SMA and bias springs

In order to transmit higher torques at low temperatures (e.g., start-up of the internal combustion engine [10]), a sliding spline sleeve joint was conceived to be externally assembled with the clutch. It is composed of two nonsliding parts (no. 5a and no. 5b), fixed to the primary and secondary shafts, respectively, and a sliding spline sleeve (no. 6) which can move under the effect of the SMA springs (no. 7) and the bias springs (no. 8).

More detail on this actuation mechanism is presented in Figure 3.14. Starting from the low temperature (about 0°C, point A), the SMA spring is in the detwinned martensite phase and the sliding mechanism is engaged. In this condition, the torque transmitted by the SMA–MR clutch exceeds 30 N·m (about 10 times greater than the engaged torque of the sole MR clutch). As the environmental temperature increases (e.g., engine start-up), the martensite to austenite transformation begins (point B, about 90°C) and the SMA spring retracts. In this condition, the equivalent tensile stress on the SMA spring increases due to the compression of the bias springs. When point C is reached, the austenite transformation is completed and the SMA spring is completely contracted, disengaging the sliding spline sleeve mechanism. In this condition, MR fluid is the main mechanism of torque transfer [12]. If the temperature is furtherly increased (point D), no transformation occurs.

When the temperature decreases, no transformation occurs up to point E, where the direct transformation begins. In the E—F path, the SMA spring elongates, as well as the bias spring, and the sliding spline sleeve joint

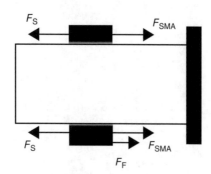

Figure 3.15 Sliding sleeve free body diagram under the effect of two bias springs force (F_C), two SMA springs force (F_{SMA}), and friction force (F_F)

engages. At point F, the martensite transformation is completed, the SMA spring is in detwinned martensite phase, and the initial configuration is restored. If the temperature furtherly decreases (path F—A), no transformation occurs.

The dynamic equation of the sliding system, schematically shown in Figure 3.15, is

$$m\ddot{\delta}_{SMA} = 2F_C(\delta_{SMA}) - 2F_{SMA}(\delta_{SMA}, \zeta) - F_F sign(\dot{\delta}_{SMA}) \qquad (3.7)$$

where m is the sliding sleeve mass, δ_{SMA} is the sleeve displacement, F_C is the bias spring force, F_{SMA} is the force applied by the SMA spring, F_F is the friction force, and ζ is the martensite fraction. It was assumed that the system is composed of two SMA springs and two bias springs. This equation along with the phase transformation and constitutive equations as presented in Chapter 1 is used to model the behavior of the system.

The springs size was chosen in order to guarantee a minimum sliding sleeve stroke of 7 mm [10]. A prototype of the sliding sleeve is shown in Figure 3.16.

A temperature profile that the SMA actuator undergoes is shown in Figure 3.17. Starting from 140°C, the temperature was decreased down to 0°C and then increased. The displacement of the device versus temperature (obtained from the model, as explained in Chapter 1) and the two steady positions of the prototype are shown in Figure 3.17.

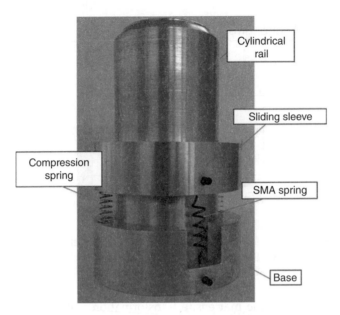

Figure 3.16 Plain dummy prototype of the sliding sleeve mechanism

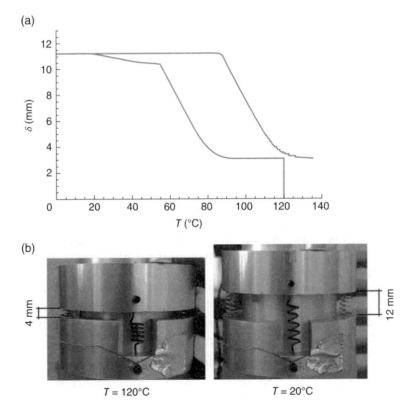

Figure 3.17 Comparison between the numerical (a) and the experimental (b) results of the sliding sleeve at different temperatures

3.4 Superelastic Actuation

Superelastic alloys are not generally considered as actuators. As shown in Figure 3.1b, these alloys undergo a large deformation in loading, which is recovered in consequent unloading. This hysteretic strain recovery can however be harnessed for actuation. In this section, three of such actuators for biomedical devices to achieve the desired shape and functionality are presented.

3.4.1 Expandable Reamer

NiTi expandable reamer aimed at cutting the bone in minimally invasive surgeries is an example of a superelastic SMA device. The outside diameter (OD) and length (L) of the blade before loading are 2.67 and 14 mm, respectively. Figures 3.18a, b, and c show the 3D, top, and side views of the reamer, respectively. This expandable device enters the body through a small incision [13]. When axially compressed by the deployment tool, the device expands and creates a four-blade structure for cutting a large channel. The OD of the blade after expansion is 8.95 mm. Once the axial load is removed, the device assumes the small-diameter shape and is removed from the body through the small incision.

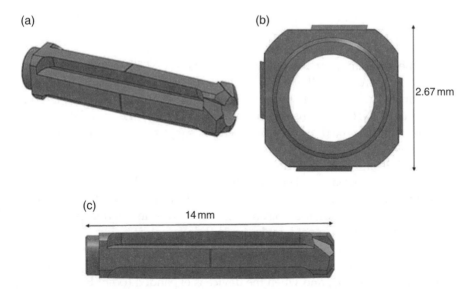

Figure 3.18 NiTi expandable reamer: (a) 3D view, (b) top view, and (c) side view

Table 3.1 Model parameters used in the simulation of the expandable reamer

Parameter	Value	Unit
E_A	38×10^3	MPa
E_M	22×10^3	MPa
M_f	−61	°C
M_s	−51	°C
A_s	−5	°C
A_f	−3.6	°C
C_M	5	Mpa/°C
C_A	6.5	Mpa/°C
v	7	—
H	0.035	—

Figure 3.19 Boundary conditions applied to the expandable reamer for the simulation

For the simulation of the blade in Abaqus, a UMAT subroutine developed for superelastic alloys [14] is used to simulate the NiTi nonlinear behaviors. The material properties of the simulated NiTi material (Fort Wayne NiTi#1) is listed in Table 3.1. During the simulation, one end of the part is fixed and the other end is subjected to 6.5 mm compression displacement (Figure 3.19).

Figure 3.20 shows the distribution of the maximum principal strain in the reamer after deformation. The NiTi device was also fabricated and experimentally evaluated. The parts were laser cut from a tube and shape set to the final shape. As shown in Figure 3.21, the blades undergo large bending deformations without considerable residual deformation.

Figure 3.22 shows the phase diagram for the superelastic transformation. Initially, the device is in austenite phase. As the stress increases, the martensite transformation occurs, and when the device is expanded (point B), the phase in certain areas is completely stress-induced martensite. When the stress is removed, the austenite phase is recovered.

EE, max. principal
(avg: 75%)

+7.883e–01
+7.228e–01
+6.572e–01
+5.917e–01
+5.261e–01
+4.606e–01
+3.950e–01
+3.295e–01
+2.640e–01
+1.984e–01
+1.329e–01
+6.732e–02
+1.778e–03

Figure 3.20 Maximum principal strain distribution in the reamer after deformation

Figure 3.21 Left, superelastic NiTi retrograde blade before deformation; right, superelastic NiTi retrograde blade after deformation. From Ref. [15]

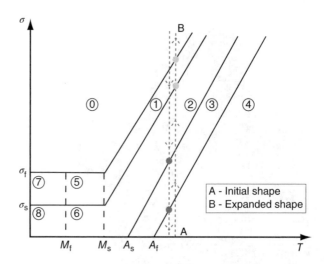

Figure 3.22 Phase diagram of the superelastic device; the dots on the transformation lines show the initial and final points of the phase transformations

3.4.2 Intervertebral Cage

The second actuator is part of an intervertebral cage for a minimally invasive spinal fusion surgery. This surgery, during which at least two adjacent vertebrae are fused together, is performed to mitigate low back pain or compression of the spinal cord [16, 17]. During the surgery, a cage, as a load-bearing implant, is permanently inserted between the two vertebrae to maintain the intermediary space. In order to be effective, these cages have a large footprint, which necessitates a large incision to insert them. To reduce the potential complications, however, it is desired to reduce the size of the incision while maintaining the large size of the cage. A cage with a superelastic actuator can change the otherwise invasive surgery to a minimally invasive endoscopic procedure.

This cage implant is a multisegment device which is completely straightened before deployment, and using the superelastic NiTi is self-configured once placed in the disc space. The cage configuration before deployment is shown in Figure 3.23. The straightened cage can be inserted through a small incision. After deployment, the cage self-configures itself and turns to a rigid oval-shaped structure (see Figure 3.24). The oval structure allows for the implant to be closer to the cortical bone walls. The implant has been developed through embedding of the NiTi hinges (via additive manufacturing) and

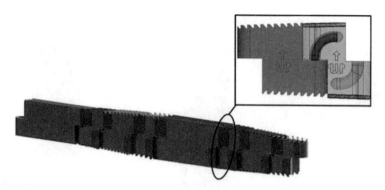

Figure 3.23   Predeployment configuration of the cage outside the body; the cage forms a line and therefore can be inserted through a small incision. From Ref. [18]

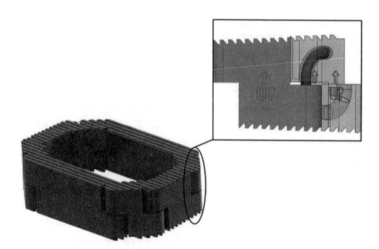

Figure 3.24   Postdeployment configuration of the cage inside the body; the superelastic hinge actuators cause the cage to form a large oval structure for added stability. From Ref. [18]

residual torque after assembly (via elliptical NiTi hinges). Also, the cage has a lordotic design to accommodate for the anatomic vertebral orientation.

To facilitate the assembly, elliptical cross section is chosen for the superelastic hinges. These hinges are fabricated by wire EDM from round superelastic bars. The cage segments are fabricated from a medical grade titanium alloy with elliptical channels to receive these hinges. The process to assemble the

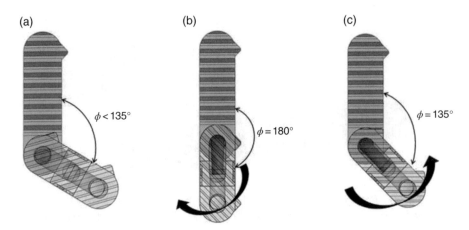

Figure 3.25 The hinge positions during multisegment assembly. One cage segment is gray; the other is transparent for clarity. The hinge is placed to connect the two segments. In panel (a), the top segment is rotated and pressed down to assume the position shown in (b). This is the predeployment form. After deployment, the cage assumes the shape shown in (c). From Ref. [18]

multisegment cage is shown in Figure 3.25. The cage segments are initially at an angle of less than 135°. It should be pointed out that the angle between the segments after deployment (inside the body) is 135° for creating the final shape of the cage. In a sequential manner, the top segments are pressed onto the bottom segments causing the elliptical hinge to follow the curved hole. As a result, the segments of the cages remain together and form an oval assembly as shown in Figure 3.26. It is worth noting that in this stage the hinges are under a minimal torque and are at the end of their elastic loading region. Before deployment, the NiTi superelastic hinge is fully loaded to the end of the superelastic plateau as the device is in the straight configuration. After deployment, the hinge unloads and the implant takes its final oval shape. Based on the required torque on segments of the cage and using simulations, the superelastic NiTi elliptical rods are designed to have the minor axis length of 1.25 mm, major axis length of 1.5 mm, and gauge length of 17.37 mm. The deployment procedure associated with the torque–angle diagram is depicted in Figure 3.26.

3.4.3 SMA Ankle–Foot Orthosis

An ankle–foot orthosis (AFO) is used to address the neuromuscular deficiencies that cause drop foot in patients who have suffered a stroke or injuries. The

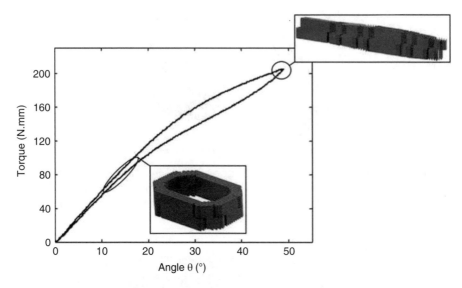

Figure 3.26 Experimental results for the NiTi hinge design with a deployment concept overlayed. The NiTi superelastic hinge is fully loaded due to the assembly method and is in its completely straight form. During deployment, the NiTi hinge unloads. Finally, the cage takes its final shape while maintaining residual torque due to the elliptical NiTi hinge design. From Ref. [16]

two main functions of this device are to stabilize the ankle to avoid unwanted collision of the foot to the ground while allowing for the ankle to rotate for ease of walking. In other words, the AFO ideally should recreate the normal torsional (rotational) stiffness of the ankle. The normal ankle stiffness has been shown experimentally to be nonlinear and hysteric. The similarity of the stiffness profile of a healthy ankle with the behavior of superelastic alloys such as NiTi is the main motivation to develop an AFO using this material. It is worth noting that the SMA-based AFO is targeted for the patients who can plantarflex (push down) their feet but have limited ability to dorsiflex (pull up) their feet.

The concept of NiTi passive AFO is depicted in Figure 3.27. The superelastic SMA element captures energy during the powered plantarflexion and releases this energy during the dorsiflexion and raises the foot. Figure 3.28 shows the use of superelastic NiTi wires to create the AFO. Wires are fixed to the brace at one end and are connected to a carriage at the other end. This carriage provides sufficient freedom by moving on a slider connected to a ball joint. Pulleys are mounted on the brace to hold the required length of wire. During

Powered plantarflexion   Dorsiflexion

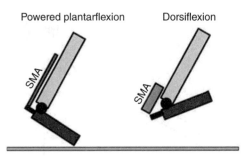

Figure 3.27 The NiTi passive AFO concept. SMA wires store energy in plantarflexion while transforming from austenite to stress-induced martensite and release it to assist during dorsiflexion while transforming back to austenite

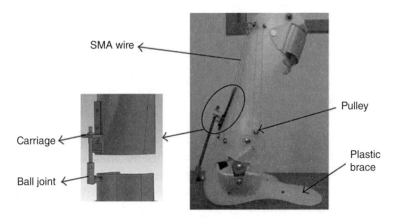

Figure 3.28 The SMA passive AFO prototype consists of two superelastic NiTi wires and an assembly to load these wires in plantarflexion and to unload in dorsiflexion. From Ref. [19]

loading and unloading, the prestrained superelastic wire's behavior is limited to the plateau section of the stress–strain curve as shown in Figure 3.1b. This is essential to keep the effort of the patients during loading of the actuator at a minimum. The length of the wire not only affects the range of motion but also determines the lifetime of the orthosis. A longer wire undergoes less strain and therefore has a longer fatigue life.

In order to evaluate the performance of the superelastic SMA AFO, gait analyses are performed on a real subject. This patient is diagnosed with left leg

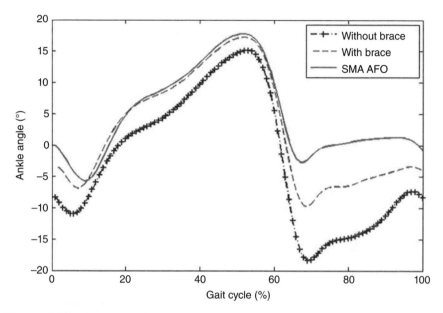

Figure 3.29 Ankle angle (mean value) versus gait percentage for three different tests

drop foot and wears a brace for his daily activities. Three sets of tests are performed: without AFO, with a hinged passive AFO, and finally with the NiTi AFO. Ankle stiffness characteristics of these three cases are compared in Figure 3.29. In absence of any AFO, there is a residual negative angle at the end of the cycle, which means that the patient is not able to raise his foot during dorsiflexion. Although wearing a brace helped him raise his foot and walk easier, there is still a large amount of residual negative angle at the end of the gait. While wearing the superelastic SMA AFO, the patient was able to dorsiflex his foot in a more desirable way.

Only one of the patient's feet is diagnosed with drop foot, so it is possible to compare the ankle stiffness characteristics of the healthy foot with the other foot while wearing the SMA AFO. These comparisons are shown in Figure 3.30a and 3.30b. Regardless of a small deviation from the desired pattern, the similarity of the behavior of the device with a healthy ankle provides the patient the ability to fully recover the moment and the angle at the end of the gait. A notable point here is that due to atrophy, muscles of the foot suffering from drop foot are not as strong as those of a healthy one, thus identical angle and moment profiles are not be expected for the two feet.

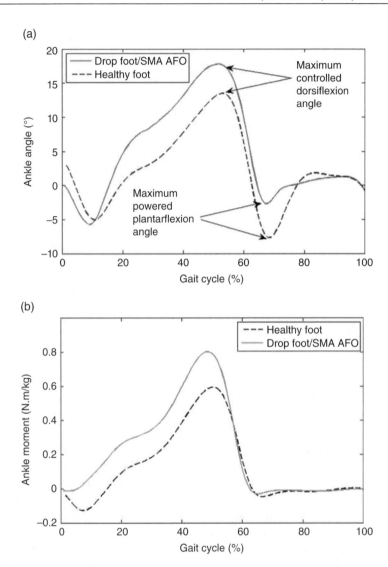

Figure 3.30 Comparison of normal angle (a) and moment (b) profiles with SMA AFO-assisted drop-foot condition

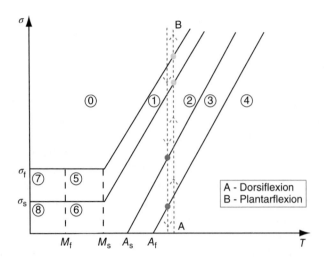

Figure 3.31 Phase diagram of the SMA ankle–foot orthosis device

Figure 3.31 shows the phase diagram for the superelastic transformation. The phase transformation behavior is similar to the one described in Figure 3.22.

3.5 Multiaxial Actuation

In majority of the current NiTi actuators, the SMA component is subjected to simple tension or pure torsion. In these simple cases, as was shown in Figure 3.1, SMA provides and/or undergoes a unique stress–strain hysteresis profile which is associated with the force/torque requirement of the device at a specific temperature. There are however certain applications, in which the required force/torque profiles are more complex than simple hysteresis loops resulted from simple tension or torsion loading/unloading. Applying combined tension–torsion loading paths is a solution to achieve and control a complicated torque–angular displacement profile. An AFO is selected and further discussed in this section to show the promise of this actuation technique.

One shortcoming of the passive SMA AFO is that the SMA stiffness in plantarflexion is higher than that in dorsiflexion. This means that the foot has to apply more force/torque to load the NiTi superelastic wire, which provides a smaller torque in unloading when it causes the foot to dorsiflex.

A solution to overcome this issue is to have an actuator undergoing both axial and torsion loading modes; one as the primary mode and the other as the secondary mode. While the primary actuation mode supplies the required force/torque of the device, the secondary mode controls the stiffness of the motion in the primary direction [20–22]. In order to show this technique, a qualitative comparison is made between the torque response of a torsion-only mode SMA actuator and a coupled axial–torsional mode SMA actuator with the same size and material. The torsion-only actuator undergoes a simple pure torsion loading/unloading. But the coupled actuator undergoes a combined out-of-phase axial and torsional loading/unloading sequence. An axial loading is first applied to the NiTi element, which causes the material to start transforming from the stiff austenite phase to the soft martensite phase prior to the rotation. Thus, the torsional stiffness is lower than a pure torsional mode, reducing the amount of torque required to rotate the actuator. When the axial load is removed in the next sequence, the SMA element partially transforms back to austenite, which raises the torsional stiffness for the next unloading block in torsion. The described loading conditions for the multiaxial loading actuator are shown in Figure 3.32, and the torque–angle responses of the two actuators are compared in Figure 3.33. Arrows in the plots show the direction of loading/unloading. In the torsion-only mode actuator, as expected from a normal superelastic NiTi, the torque of the loading plateau is higher than the torque in unloading. But in the coupled mode actuator, the torsional loading occurs at a lower torque than the unloading. This enables a robust control of the torsional response of the actuator for various applications. Similarly, by applying torsional load, it is possible to control and adjust the axial response of a superelastic element.

A proposed application for the combined mode SMA actuator in the form of an AFO actuator is shown in Figure 3.34. An SMA rod actuator is fixed to the base at one end and is coupled with a linear actuator at the other end. The linear actuator rotates with the rotation of an arm connected to the bottom of the device and applies axial displacement to the SMA rod when necessary.

Step 1   Step 2   Step 3   Step 4

Figure 3.32 Four-step multiaxial loading of a NiTi rod consists of 1 axial loading, 2 torsional loading, 3 axial unloading, and 4 torsional unloading.

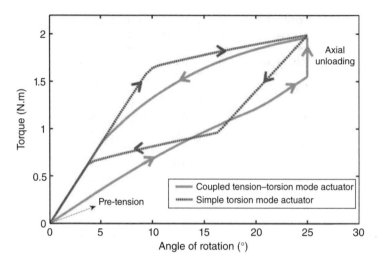

Figure 3.33 Comparison of a passive torsion mode actuator and an active coupled mode actuator

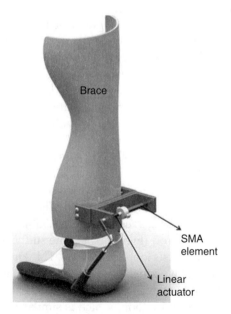

Figure 3.34 A multiaxial SMA AFO

Table 3.2 Model parameters used for the
numerical simulation of SMA AFO

Parameter	Value	Unit
E_A	20.0×10^3	MPa
E_M	10.0×10^3	MPa
M_f	-32.0	°C
M_s	-15.0	°C
A_s	0	°C
A_f	10.0	°C
C_M	8.3	MPa/°C
C_A	7.5	MPa/°C
v	0.33	—
H	0.03	—

The torque–angle curves of the human ankle joint have been extensively studied by conducting motion analysis on able-bodied subjects [23]. The torque–angle response (angular stiffness) changes as the walking speed increases. The loading and unloading portions of the moment versus angle curves show clockwise loop (hysteresis) at slow speeds that reduces as the speed increases to normal speeds. Above the normal speeds, the loops start to traverse to a counter clockwise path. To mimic this behavior, a system with active components and complex sensing and control strategies would be necessary.

As an alternative solution, the coupled actuation method is used to mimic the response of the ankle and thereby providing assistance to drop-foot patients. The SMA rod has the diameter of 3.5 mm and length of 10 mm at the temperature of 20°C in which the NiTi element is initially in its austenitic phase. Material properties are listed in Table 3.2.

The moment versus angle curve of an SMA rod shows a clockwise hysteresis loop when it is subjected to a pure torsion (Figure 3.35a). This response corresponds to the slow walking speeds (Figure 3.35b). As an extra axial load is being applied to the NiTi rod (Figure 3.36a, the stiffness of the component changes such that the hysteresis loop disappears. This kind of behavior would be correlated to the normal walking speed (Figure 3.36b). Finally, at high axial loads (Figure 3.37a), the hysteresis loop of the moment versus angle curve approaches a counter clockwise loop, which is similar to the ankle behavior at high walking speeds (Figure 3.37b) [24].

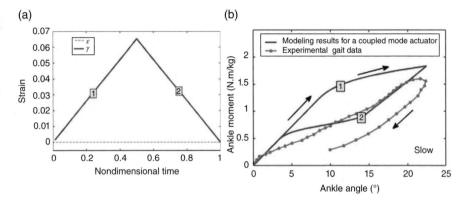

Figure 3.35 An active SMA AFO in combined axial–torsional mode at the slow walking speed. (a) Strain history applied to the SMA; the SMA component is subject to a pure torsion, (b) Comparison of the experimental data [23] against the modeling results of the coupled mode actuator for the ankle moment versus angle curves at slow walking speed (1.2 m/s)

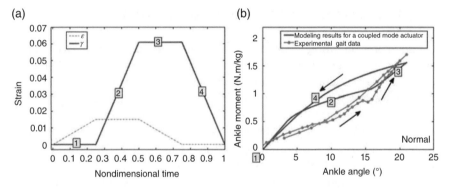

Figure 3.36 An active SMA AFO in combined axial–torsional mode at the normal walking speed. (a) Strain history applied to the SMA. SMA component is subject to torsion and a small axial strain of 1.5%, (b) Comparison of the experimental data [23] against the modeling results of the coupled mode actuator for the ankle moment versus angle curves at normal walking speed (1.5 m/s)

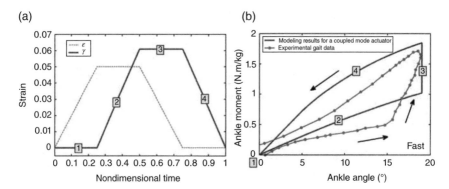

Figure 3.37 An active SMA AFO in combined axial-torsional mode at the fast walking speed. (a) Strain history applied to the SMA. SMA component is subject to torsion and the axial strain of 5%, (b) Comparison of the experimental data [23] against the modeling results of the coupled mode actuator for the ankle moment versus angle curves at fast walking speed (1.9 m/s)

3.6 Antagonistic SM-Superelastic Actuation

An interesting actuation mechanism is possible by employing the two properties of SMAs: SM effect and superelasticity (SE). An SM member when combined with an SE element creates an antagonistic actuator [13, 25]. The SM segment controls the actuation function, while the system is designed and operated in a way that the SE element provides the opposing actuation force and stroke. Let's assume that the actuator works between two temperatures: low temperature (T_l) and high temperature (T_h). The SE member is always in the austenite phase; that is its austenite finish temperature is lower than both the low and high temperatures ($A_f^{SE} < T_l, A_f^{SE} < T_h$). On the other hand, the SM element is initially in its martensite phase, which transforms to austenite at the high temperature. To this end, the SM material should be selected such that its martensite finish temperature (M_f^{SM}) is below the low temperature and its austenite finish temperature (A_f^{SM}) below the high temperature (i.e., $T_l < M_f^{SM}$ and $T_h > A_f^{SM}$). The geometry of the SE and SM components and their memorized shape are set such that in low temperature, the memorized shape of the SE element is stronger and the assembly moves toward the SE element, which is designed to be the low-temperature (inactive) form of the actuator. At the high temperature, the SM segment exceeds its austenite temperature and deflects the assembly toward its memorized shape that is designed to be the high temperature (active) configuration of the device.

3.6.1 Cardiac Clamp

The described antagonistic actuator could be developed in several different ways. A possible configuration is based on the SM element as a flat wire and the SE element as a ring shaped round wire. By wrapping the flat wire around the round SE member, an antagonistic actuator is formed.

When the two members are assembled at a low temperature, they apply force on each other in a way that the SM wire tends to close the ring, while the SE element tends to keep its memorized open ring shape. Depending on the material properties, geometry, and the memorized shape of each of the two components, an equilibrium position will be obtained at the low temperature. Figure 3.38a displays the equilibrium or neutral position of the device at low temperature. By heating the SM element either by changing the thermal environmental condition or by resistive heating, the assembly moves toward the shape set form of the SM member, which tends to close the ring as shown in Figure 3.38b. By cyclic heating and cooling, these two configurations can be repeated. It is worth mentioning that the actuation stroke and shape generally depend on the geometry, material properties, and shape set forms of the SM and SE elements; thus, numerous actuation

(a)   (b)

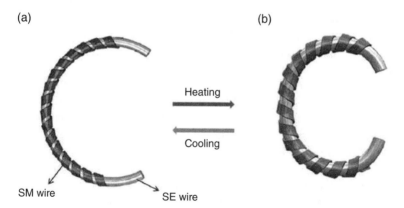

Heating

Cooling

SM wire   SE wire

Figure 3.38 An antagonistic superelastic wire and shape memory flat wire wrapping creates a bistable actuation mechanism. (a) Low-temperature (left) and (b) high-temperature stable forms (right) provide the close and open forms of the actuator. Courtesy of Lifewire LLC Macon GA

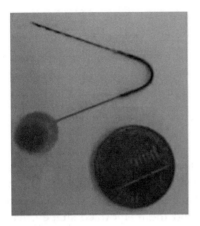 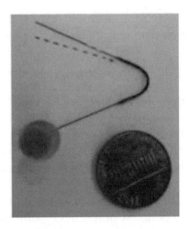

Figure 3.39 An antagonistic superelastic flat wire wrapped around a shape memory wire creates a two-way actuation with repeatable actuation between two stable positions: low-temperature (left) and high-temperature (right) positions

configurations are possible with the antagonistic actuator design. Figure 3.39 shows a similar configuration where an actuator is created from a flat superelastic wire, which is wrapped around a SM wire to create a bidirectional bistable actuation.

Various clamps are used to hold the tissues and organs during a surgery. The conventional devices are usually large, heavy, and hard to control, especially when the surgery is minimally invasive with limited visibility and mechanical access. An SMA cardiac tissue clamp is designed using the discussed antagonistic actuation mechanism. This device is light and small, with a large actuation stroke and the possibility of remotely controllable releasing.

Even though finite element is a common approach for modeling SMA actuators, modeling contact and behavior of multiple elements using this methodology is a complex problem. In the following, an alternative method of equivalent stiffness for design and analysis of these actuators is presented in the context of designing the cardiac clamp.

The design process starts by assuming an equivalent representative geometry to model the behavior of the wrapped flat wire. To this end, the flat wire is modeled as a tube with the same radius and thickness. As shown in Figure 3.40, the two elements are shape set in the desired configurations that they should assume. When the two elements are in contact in the final assembly, the antagonistic actuator assumes a neutral position as shown in Figure 3.40c.

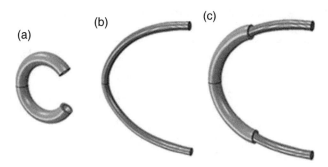

Figure 3.40 The wrapped flat shape memory wire actuator is modeled as a tube (a) shape set form of the tube and (b) shape set form of the superelastic antagonistic element; (c) when assembled, each of the two elements undergoes a deformation, and the assembly finds an equilibrium position

To design the geometry of the two antagonistic elements, the approach is based on creating the desired level of stiffness in each element. In the case of the tissue clamp, flexural stiffness (rigidity) is the basis for the design process. For other actuation mechanisms, the appropriate stiffness is used for the design process. Stiffness of a tube with inner diameter of d_i and an outer diameter of d_o can be written as

$$k_{tube} = \frac{\pi\left(d_o^4 - d_i^4\right)E}{64} \tag{3.8}$$

A more realistic representation of the behavior of the flat wire, which forms a helical wrap, can be achieved by using the lateral bending stiffness of a helical spring with rectangular cross section [26]. The lateral rigidity for a relatively closed-coiled helical spring with circular cross section and L, d, n, D, E, and G as the length, wire diameter, number of active coils, mean coil diameter, Young's modulus, and shear modulus, respectively, can be written as-

$$k_{cir} = \frac{ELd^4}{32nD(1 + (E/2G))} = \frac{2LI_{cir}}{n\pi}\left(\frac{E}{D(1 + (E/2G))}\right) \tag{3.9}$$

where $I_{cir} = \pi d^4/64$. By replacing I_{cir} with $I_{rec} = bh^3/12$, the relation for the rigidity of the rectangular cross section is

$$k_{flat} = \frac{bh^3 L}{6\pi n}\frac{E}{D(1 + (E/2G))} \tag{3.10}$$

Table 3.3 Model parameters used for the numerical simulation of SMA cardiac clamp

Parameter	SE	SM	Unit
E_A	27.0×10^3	36.0×10^3	MPa
E_M	20.0×10^3	19.0×10^3	MPa
M_f	27.4	−51.9	°C
M_s	35.5	−48.7	°C
A_s	47.6	7.5	°C
A_f	54.2	19.4	°C
C_M	8	7	MPa/°C
C_A	8	4.4	MPa/°C
v	0.33	0.33	—
H	0.06	0.052	—

By replacing $n = L/b$ and $G = E/2(1 + v)$, where v is the Poisson's ratio and assumed to be 0.33 for NiTi, the final approximate expression for the lateral rigidity of the NiTi flat wire is

$$k_{flat} = \frac{b^2 h^3 E}{44D} \qquad (3.11)$$

In Equation 3.11, b and h are the dimensions of the flat wire SM wire, and D is the diameter of the round superelastic wire. For the material properties, parameters listed in Table 3.3 are employed. The stiffness from Equation 3.11 is then used to find an equivalent flat wire actuator to reproduce the tube stiffness of Equation 3.8. At high temperatures, the SM element transforms to austenite. Recalling that $E_A > E_M$ means the flat wire will become stiffer. This higher stiffness should be sufficient to deform the superelastic round wire. In other words, at higher temperatures, the stiffness of the SM flat wire should be greater than the stiffness of the round SE wire. On the other hand, at the low temperature, the stiffness of the round SE wire should be greater than the stiffness of the SM wire.

A cardiac tissue clamp is designed using this approach [13]. A round SE wire with diameter of $D = 0.381$ mm is used with a flat wire with $h = 0.2$ mm and $b = 1$ mm. The device delivers a tip displacement of 2 mm. A prototype of the device is then fabricated and experimentally evaluated. Figure 3.41a shows the device before the actuation of the SM wire at room temperature. In this state, the clamp is in the closed (engaged) configuration to secure the tissue. As soon as the SM wire is heated, it starts to actuate and disengage the

(a)   (b)

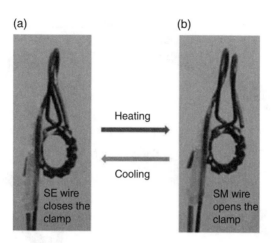

Heating

Cooling

SE wire closes the clamp

SM wire opens the clamp

Figure 3.41 The superelastic cardiac tissue clamp is disengaged by heating the shape memory flat wire. (a) Before actuation, the clamp is in the engaged position. (b) After actuation, the clamp is in the disengaged position

clamp as shown in Figure 3.41b. By cyclic heating and cooling, the clamping and releasing configurations are repeated.

3.6.2 Expandable Pedicle Screw

In general, a superelastic element generates a more favorable bias force with an initial linear force–displacement profile followed by a close-to-zero-stiffness plateau, as was shown in Figure 3.1. The main advantage of this bias form is that an actuator made of a SM actuator and a superelastic element is more efficient in assuming various shapes. This type of actuator has been employed as the anchoring elements to improve bone-implant integrity [25, 27] as shown in Figure 3.42. In the resulting pedicle screw, the added antagonistic SM–SE actuators have two stable forms: retracted and expanded as shown in Figure 3.42b and d, respectively. The superelastic tube is shape set to the retracted form, and the SM core is shape set to the expanded form. In the low-temperature form, as shown in Figure 3.42a, the SM element is in temperature-induced martensite, and the superelastic element retracts the actuators to the pedicle screw as shown in Figure 3.42b. At the body temperature, which is above the austenite final of the SM element, the second stable form—the expanded form of the screw—is achieved, as shown in Figure 3.42c and d. As the screw assembly is cooled below the martensite final temperature of the SM element, the superelastic element moves

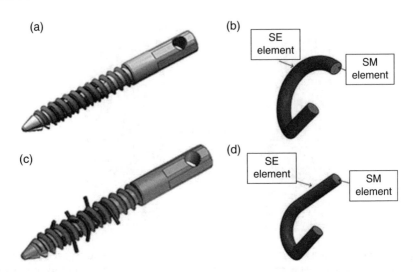

Figure 3.42 A pedicle screw with a series of combined shape memory (SM) and superelastic (SE) actuators; (a) at low temperature before implanting, (c) at body temperature after implant, (b) retracted actuator at low temperature, and (d) expanded actuator at body temperature [25]

the actuator assembly back to the retracted configuration. The resulting bone screw has the advantage of mitigating osteoporosity by maintaining the long-term anchoring force as the surrounding bone degrades. This design has a superior pullout strength in osteoporotic bones, which is demonstrated using FEA [25].

Figure 3.43 shows the phase diagram for the combined SM–SE device. The dotted lines show the SM and SE paths. Points A and B indicate the two stable shapes of the device (A is relaxed; B is contracted). The light gray points identify the beginning and the end of martensite transformation, and the dark gray points pinpoint the start and finish of austenite transformation.

3.7 Summary

In this chapter, different actuation mechanisms of SM alloys are discussed and their applications are introduced. Design methodology is qualitatively explained for each application, and the evaluation methodology is presented when available. SM alloys provide various forms of actuation in both SM and superelastic regimes. SM actuators such as the SMA rotary system (Figure 3.2)

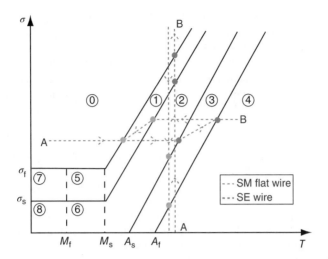

Figure 3.43 Phase diagram combined SM–SE devices. The black dots indicate the heating transformation, and black squares indicate the cooling transformation for both SM and SE alloys. Reproduced with permission from Ref. [25], Hindawi Publishing Corporation

can produce repetitive motions with minimal number of system components. Antagonistic SM–SM systems like the SMA mirror (Figure 3.7) can deliver the more stringent actuation requirements. Superelastic actuators are fast, passive, precise, and robust. They can be used in different geometrical forms and applications such as the expandable reamer (Figure 3.21) or the elliptical rod in an intervertebral cage (Figure 3.25). Superelastic actuators may be used under simple loading conditions like the AFO in Figure 3.28 or more complex combined loading conditions such as the multiaxial AFO in Figure 3.34. Antagonistic SM–SE actuators such as the cardiac tissue (Figure 3.41) benefit from the controllable characteristic of the SM element, while the superelastic component provides superior force and stroke.

References

[1] Elahinia M, Ashrafiuon H, Ahmadian M, Tan H. A temperature-based controller for a shape memory alloy actuator. Journal of Vibration and Acoustics. 2005; **127**(3):285–291.

[2] Elahinia M, Ashrafiuon H. Nonlinear control of a shape memory alloy actuated manipulator. Journal of Vibration and Acoustics. 2002;**124**(4):566–575.

[3] Elahinia M, Seigler T, Leo D, Ahmadian M. Nonlinear stress-based control of a rotary SMA-actuated manipulator. Journal of Intelligent Material Systems and Structures. 2004;**15**(6):495–508.

[4] Elahinia M, Ahmadian M, Ashrafiuon H. Design of a Kalman filter for rotary shape memory alloy actuators. Smart Materials and Structures. 2004;**13**(4):691.

[5] Elahinia M, Koo J, Ahmadian M, Woolsey C. Backstepping control of a shape memory alloy actuated robotic arm. Journal of Vibration and Control. 2005; **11**(3):407–429.

[6] Williams E, Elahinia M. An automotive SMA mirror actuator: modeling, design, and experimental evaluation. Journal of Intelligent Material Systems and Structures. 2008;**19**(12):1425–1434.

[7] Williams E, Shaw G, Elahinia M. Control of an automotive shape memory alloy mirror actuator. Mechatronics. 2010;**20**(5):527–534.

[8] Bucchi F, Elahinia M, Forte P, Frendo F. A passive magneto-thermo-mechanical coupling actuated by SMA springs and MR fluid. International Journal of Structural Stability and Dynamics. 2014;**14**(8):1440031.

[9] McNaney JM, Imbeni V, Jung Y, Papadopoulos P, Ritchie R. An experimental study of the superelastic effect in a shape-memory Nitinol alloy under biaxial loading. Mechanics of Materials. 2003;**35**(10):969–986.

[10] Bucchi F, Elahinia M, Forte P, Frendo F. Development and testing of a hybrid SMA-MR passive clutch. Proceedings of the ASME 2014 Smart Materials, Adaptive Structures and Intelligent Systems, SMASIS Newport, RI, USA. September 8–10, 2014.

[11] Bucchi F, Forte P, Frendo F. A fail-safe magnetorheological clutch excited by permanent magnets for the disengagement of automotive auxiliaries. Journal of Intelligent Material Systems and Structures. 2014;**25**(16):2102–2114.

[12] Bucchi F, Forte P, Frendo F, Squarcini R. A magnetorheological clutch for efficient automotive auxiliary device actuation. Frattura ed Integrità Strutturale. 2013;**23**:62–74.

[13] Andani MT, Elahinia M. Modeling and simulation of SMA medical devices undergoing complex thermo-mechanical loadings. Journal of Materials Engineering and Performance. 2014;**23**(7):2574–2583.

[14] Lagoudas D, Bo Z, Qidwai M, Entchev P. SMA UM: user material subroutine for thermomechanical constitutive model of shape memory alloys. Texas A&M University College Station, TX; 2003.

[15] Elahinia, M. Taheri Andani, M. and Haberland, C. "Shape memory and superelastic alloys," in High Temperature Materials and Mechanisms, Editor: Yoseph Bar-Cohen, CRC Press/Taylor and Francis Group, FL, 2014, ISBN 9781466566453.

[16] Andani MT, Anderson W, Elahinia M. Design, modeling and experimental evaluation of a minimally invasive cage for spinal fusion surgery utilizing superelastic Nitinol hinges. Journal of Intelligent Material Systems and Structures. 2014;**26**(6) 631–638.

[17] Anderson W, Chapman C, Karbaschi Z, Elahinia M, Goel V. Design and testing of a minimally invasive intervertebral cage for spinal fusion surgery. Smart Structures and Systems. 2013;**11**(3):283–297.

[18] Anderson W. Development of an intervertebral cage using additive manufacturing with embedded NiTi hinges for a minimally invasive deployment. University of Toledo, Toledo, OH; 2013.

[19] Deberg L, Andani MT, Hosseinipour M, Elahinia M. An SMA passive ankle foot orthosis: design, modeling, and experimental evaluation. Smart Materials Research. 2014;**2014**:1–11.

[20] Andani MT, Alipour A, Eshghinejad A, Elahinia M. Modifying the torque–angle behavior of rotary shape memory alloy actuators through axial loading: a semi-analytical study of combined tension–torsion behavior. Journal of Intelligent Material Systems and Structures. 2013;**24**(12):1524–1535.

[21] Andani MT, Elahinia M. A rate dependent tension–torsion constitutive model for superelastic nitinol under non-proportional loading; a departure from von Mises equivalency. Smart Materials and Structures. 2014;**23**(1):015012.

[22] Andani MT, Alipour A, Elahinia M. Coupled rate-dependent superelastic behavior of shape memory alloy bars induced by combined axial-torsional loading: a semi-analytic modeling. Journal of Intelligent Material Systems and Structures. 2013;**24**(16):1995–2007.

[23] Hansen A, Childress D, Miff S, Gard S, Mesplay K. The human ankle during walking: implications for design of biomimetic ankle prostheses. Journal of Biomechanics. 2004;**37**(10):1467–1474.

[24] Gorzin Mataee M, Andani MT, Elahinia M. Adaptive ankle foot orthoses based on superelasticity of shape memory alloys. Journal of Intelligent Material Systems and Structures. 2014;**26**(6):639–651.

[25] Eshghinejad A, Elahinia M, Goel V. Functionality evaluation of a novel smart expandable pedicle screw to mitigate osteoporosis effect in bone fixation: modeling and experimentation. Smart Materials Research. 2013;**2013**:1–8.

[26] Chironis N. Spring design and application, vol. **276**. McGraw-Hill, New York; 1961.

[27] Tabesh M, Goel V, Elahinia M. Shape memory alloy expandable pedicle screw to enhance fixation in osteoporotic bone: primary design and finite element simulation. Journal of Medical Devices. 2012;**6**(3):034501.

4

Control of SMA Actuators

Hashem Ashrafiuon and Mohammad H. Elahinia

Control is an integral part of SMA actuation. This chapter highlights the unique challenges in designing control algorithms for these systems. Issues such as interdependency of the dynamics of the system and behavior of the actuator will be discussed. Dynamics of the system under actuation defines the stress level on the SMA element providing the actuation force. As the stress level changes during the motion, the phase transformation properties of the actuator evolve. The control system therefore should mitigate this effect for effective regulation of the behavior of the system. Position control of SMA actuator using nonlinear controllers is discussed in this chapter where we concentrate on sliding mode control since it is robust to uncertainties and disturbances.

Due to nonlinear hysteresis behavior of SMAs, the control of SMA actuators is a challenging task. The hysteresis properties existing in SMA involve the relationships between displacement and temperature, Young's modulus and temperature, displacement and voltage, strain and temperature as well as martensite fraction and temperature. Another issue in controlling SMA is its challenging actuation impetus (adding and removing heat). The SMA actuation is mostly initiated by applying a current to the element which

Shape Memory Alloy Actuators: Design, Fabrication, and Experimental Evaluation, First Edition. Mohammad H. Elahinia.
© 2016 John Wiley & Sons, Ltd. Published 2016 by John Wiley & Sons, Ltd.

is therefore subject to Joule heating and natural convection. The SMA system may experience a large wait time for the element to cool. Expedited heat removal techniques are usually in the form of using heat sink, water immersion, and/or forced convection. The temperature and type of environment, the convection of the environment, and the surface to volume ratio of the SMA elements could all influence the heating and cooling time. These time delays for SMA actuators to respond to control signals add an additional complexity to the nonlinear problem and present a challenge in controller design.

In the design of an SMA controller, it is important to avoid overheating and/or overloading. If an SMA component is overheated, it will prolong the cooling procedure, decreasing the speed of the actuation. While not contributing to the actuation, overloading might cause a malfunction or decrease the fatigue life of the system.

Several methods of controlling SMA actuators have been investigated; most of these control systems, however, are nonmodel based with limited possibility for stability and robustness analysis. Different variations of linear proportional–integral–derivative (PID) controls have been explored [1–5], while the earlier designs used pulse width modulation [6, 7]. Several nonlinear control schemes such as fuzzy logic, neural networks, feedback linearization, sliding mode control, and variable structure control have also been explored by researchers [8–12]. It is also possible to combine these control techniques, for example, by using sliding mode feedback control in conjunction with a feedforward neural network [13]. Neural networks have shown great possibility in mapping and adaptation for capturing the hysteresis behavior of the SMA. However, neural networks need training to operate and therefore the training data and training methods are critical. This may require a large amount of data to secure the statistical accuracy. Furthermore, other controllers have also been designed that include an additional open-loop part for improving the performance [14, 15]. A model-based backstepping controller has been applied for the SMA actuated robotic arm that was introduced in Chapter 1 [16]. The system has been the subject of a stress-based and temperature-based controller which demonstrated enhanced tracking performance for rotary SMA actuators [17, 18]. These efforts have been complemented by designing a general sliding mode control law for mechanical systems actuated by SMA [19]. A comprehensive review of control methods using SMA actuators is presented in [20].

In this chapter, we will provide an overview of evolution of sliding mode control methods which were applied to robotics arm actuated by SMA. These methods start with simple nonmodel-based control laws and end with more sophisticated model-based control laws that can be applied to generic mechanical systems actuated by SMA. We will also discuss the variables that can be measured and those that must be estimated.

4.1 Introduction to Sliding Mode Control

The most unique feature of sliding mode control is its ability to result in very robust control systems which are insensitive to parametric uncertainty and external disturbances [21–23]. Here, we provide a short overview of the method to prepare the reader for the application of sliding mode control to SMA actuators.

Consider a single-input dynamic system:

$$x^n = f(x) + b(x)u \tag{4.1}$$

where the scalar x is the output, scalar u is the control input, $x = \left[x\,\dot{x}...x^{n-1}\right]^T$ is the state vector, and n is the number of states. In the above equation, the drift term $f(x)$ and control gain $b(x)$ are not exactly known, but both assumed to be bounded by known, continuous functions of x. The control problem is to get the state x to track a specific time-varying state $x_d = \left[x_d\,\dot{x}_d...x_d^{n-1}\right]^T$ in the presence of modeling uncertainties mentioned previously.

Define the tracking error in the state x as

$$\tilde{x} = \left[\tilde{x}\,\dot{\tilde{x}}...\tilde{x}^{n-1}\right]^T \tag{4.2}$$

where $\tilde{x} = x - x_d$ is the state vector tracking error. Next, define a single time-varying nth-order surface as

$$s = \left(\frac{d}{dt} + \lambda\right)^{n-1}\tilde{x} \tag{4.3}$$

where λ is a strictly positive constant. For instance, if $n = 2$, then the surface is given as $s = \dot{\tilde{x}} + \lambda\tilde{x}$.

For system to track the desired state, it can be seen that time-varying surface $s(t)$ is equal to zero, which represents a linear differential equation problem whose solution exponentially approaches zero. Hence, the problem of tracking the n-dimensional vector x_d can in effect be replaced by a first-order stabilization problem in $s(t)$. The control derivation is then as follows.

We differentiate $s(t)$ and set it equal to zero to obtain a nominal (or equivalent) control law. Furthermore, any bounds on $s(t)$ can be directly translated into bounds on the tracking error vector $\tilde{x}$, and therefore the scalar s represents a true measure of tracking performance. Next, we define the Lyapunov candidate function $\frac{1}{2}s^2$ and choose the control law such that its time derivative is negative definite function:

$$\frac{1}{2}\frac{d}{dt}s^2 \leq -\eta|s| \tag{4.4}$$

where η is a strictly positive constant. The aforementioned inequality is called the reaching condition in sliding mode control since it guarantees all trajectories reaching the surface in finite time. Once the system trajectory reaches the surface, then it will remain on the surface and slides to the origin since the surface as defined in (4.3) is asymptotically stable. The control is normally defined by subtracting a signum function with a positive gain from the nominal control.

Let us consider the second-order case where input gain is unity (i.e., $b(x) = 1$) to illustrate a simple illustration of the control law derivation:

$$\ddot{x} = f(x) + u \tag{4.5}$$

where f is estimated as $\hat{f}$ and is bounded by a known function $F = F(x, \dot{x})$:

$$\left|\hat{f} - f\right| \leq F \tag{4.6}$$

Defining the surface as $s = \dot{\tilde{x}} + \lambda\tilde{x}$ and setting $\dot{s} = 0$, the nominal control law is derived as

$$\tilde{u} = -\hat{f} + \ddot{x}_d - \lambda\dot{\tilde{x}} \tag{4.7}$$

and the sliding mode control law is given by

$$u = \hat{u} - b\,\mathrm{sgn}(s) \tag{4.8}$$

where "sgn" is the signum function, such that

$$\mathrm{sgn}(s) = \begin{cases} +1 & \text{if } s > 0 \\ 0 & \text{if } s = 0 \\ -1 & \text{if } s < 0 \end{cases} \tag{4.9}$$

and k must be selected to satisfy the reaching condition

$$\frac{1}{2}\frac{d}{dt}s^2 = \dot{s}s = \left(f-\hat{f}\right)s-k|s| \le -\eta s$$

by choosing $k = F + \eta$. Furthermore, the discontinuous signum function normally results in chattering which may be avoided by a continuous approximation such as a hyper tangent function, $\mathrm{sgn}(s) \approx \tanh(s/\phi)$, where ϕ can be increased to suppress chattering, but the resulting control law is only guaranteed to converge within a "small" region near the origin. The simple procedure explained earlier is followed in all the examples of this chapter.

4.2 Sliding Mode Control of SMA Actuators without Modeling

Grant and Hayward [11] used a variable structure control (sliding mode control) for an SMA camera actuator comprised of a pair of antagonist actuators. They developed two-stage linear and two-stage constant magnitude controllers. These two controllers apply electric current based upon the position error. No model was used in the control law. The sliding surface was defined as the position error. Hence, only position feedback is required for this control law, and the control switches back and forth between the two antagonist actuators. They were able to show that the two-stage constant magnitude controller is more robust and at the same time has a simpler design.

A more interesting control problem is when only a one-way SMA actuator is employed per degree of freedom (DOF) in a mechanical system. Elahinia and Ashrafiuon [24] introduced a nonmodel-based sliding mode control law for a one-DOF rotary robotic arm actuated by SMA wire and pulley system (introduced in Chapter 1). Figure 4.1 shows a picture and a drawing of the arm which uses a bias spring to aid with the reverse rotation. The properties of the robot arm and the SMA actuator are listed in Table 4.1. As was shown in Chapter 1, the system between the input voltage and output angular position is of order 3. This third-order dynamics is comprised of the first-order heat transfer and the second-order rotary motion (dynamics) of the arm. A simple switching sliding mode control law can be designed based on position feedback only. In this control law, the surface is of zero order and may be defined as the error in joint angle:

$$s = \tilde{\theta} = \theta - \theta_\mathrm{d} \tag{4.10}$$

where s is the surface, θ is the joint angle, θ_d is the desired joint angle, and $\tilde{\theta}$ is the position error. The control law is then defined based on the limitations of

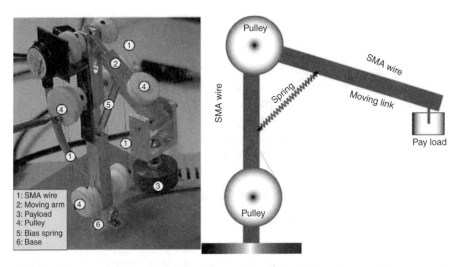

Figure 4.1 A one-degree-of-freedom rotary SMA NiTi actuator; the upward motion is generated by shape memory effect and the downward motion by the bias spring and the weight of the payload. Reproduced with permission from Ref. [24], ASME

the actuator and hardware to switch between a high and a low voltage values similar to (Grant and Hayward [11]):

$$u = \begin{cases} V_{\text{high}} & s < 0 \\ 0 & s = 0 \\ V_{\text{low}} & s > 0 \end{cases} \tag{4.11}$$

Since the control law resulted in chattering, it was then modified by using a saturation function as follows:

$$u = \begin{cases} V_{\text{high}} & \dfrac{s}{\phi} < -1 \\ V_{\text{low}} + \left(V_{\text{high}} - V_{\text{low}} \right) s/\phi & -1 \leq \dfrac{s}{\phi} \leq 1 \\ V_{\text{low}} & \dfrac{s}{\phi} > 1 \end{cases} \tag{4.12}$$

Next, the surface was modified to a first-order surface to improve the controller performance:

$$s = \dot{\tilde{\theta}} + \lambda \tilde{\theta} \tag{4.13}$$

Table 4.1 Parameters of the one-degree-of-freedom SMA rotary actuator

Parameters	Description	Unit	Value
m	SMA wire's mass per unit length	kg	$1.414e^{-4}$
A	SMA wire's circumferential area per unit length	m^2	$4.712e^{-4}$
C	Specific heat of wire	kcal/kg·°C	0.2
R	SMA wire's resistance per unit length	Ω	45
T_∞	Ambient temperature	°C	20
H	Heat convection coefficient	$J/(m^2 \cdot °C \cdot s)$	150
E_A	Austenite Young's modulus	GPa	75.0
E_M	Martensite Young's modulus	GPa	28.0
θ_T	SMA wire's thermal expansion factor	MPa/°C	0.55
Ω	Phase transformation contribution factor	GPa	−1.12
σ_0	SMA wire's initial stress	MPa	75.0
ε_0	SMA wire's initial strain		0.04
T_0	SMA wire's initial temperature	°C	20
ξ_0	SMA wire's initial martensite fraction factor		1.0
A_s	Austenite start temperature	°C	68
A_f	Austenite final temperature	°C	78
M_s	Martensite start temperature	°C	52
M_f	Martensite final temperature	°C	42
C_A	Effect of stress on austenite temperatures	MPa/°C	10.3
C_M	Effect of stress on temperatures	MPa/°C	10.3
L_0	Initial length of SMA wire	mm	900
R	Pulley diameter	mm	8.25
m_p	Pay load mass	gr	57.19
m_a	Moving link mass	gr	18.7
K	Bias spring stiffness	N/m	3.871

The resulting control law, however, required angular velocity feedback which was approximated numerically using the position feedback measurements.

Since the control law is not model based, the "optimum" values for the control parameters must be derived through simulations and experiments. We chose these values to be $\phi = 10°$ and $\lambda = 10$. The evolutionary performance improvement of the three control laws can be verified through simulations. Figure 4.2 shows the performance of the discontinuous zero-order (1), continuous zero-order (2), and continuous first-order (3) methods. Since the superior performance of the first-order surface is established, experiments can be performed with the control law based on the first-order surface (3). The set-point

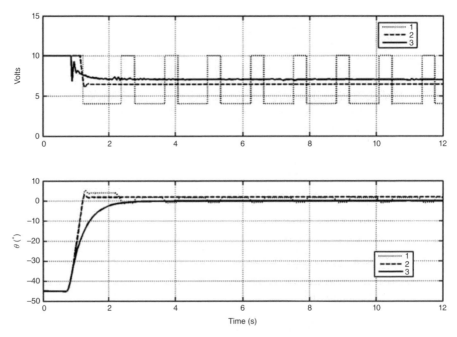

Figure 4.2 Control input V and output θ using the three control laws: discontinuous zero order (1), continuous zero order (2), and continuous control based on first-order surface (3)

control performance was shown to be robust for a variety of desired angle, as shown in Figure 4.3. However, there tracking errors were significant when the actuator was commanded to track a sinusoidal trajectory with a 1 rad/s frequency, as shown in Figure 4.4.

This work was continued by adding integral sliding mode control law and applied to a three-DOF rotary robotic arm where the second and third joints were actuated by SMA wire and pulley systems [25]. The robotic arm is shown in Figure 4.5, and the experimental control system is diagramed in Figure 4.6. The setup consists of a dSPACE board and panel, dSPACE software installed on a PC, two programmable power supplies, the robot arm, two position encoders, two digital to analog converters, and an EyeBot board to control the servomotor. The complete set of parameter values for the robot are summarized in Tables 4.2, 4.3, 4.4, and 4.5. These parameter values will be relevant for the model-based controller presented later in this chapter.

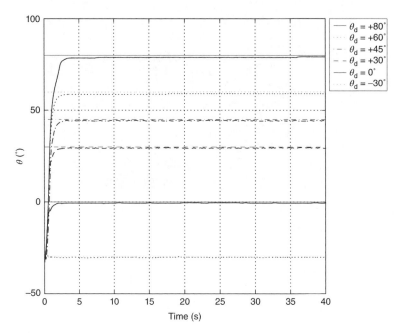

Figure 4.3   Experimental results with modified for different desired angles

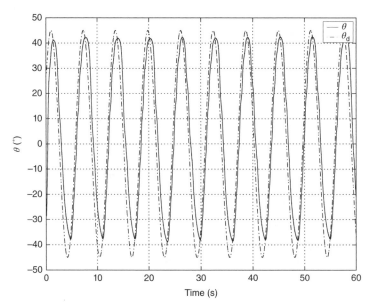

Figure 4.4   Experimental results with VSC3 for a sinusoidal trajectory tracking

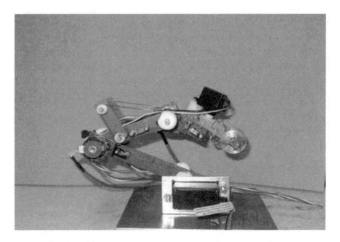

Figure 4.5 The 3DOF SMA actuated robot arm carrying a load; two degrees of freedom are actuated by SMA wires

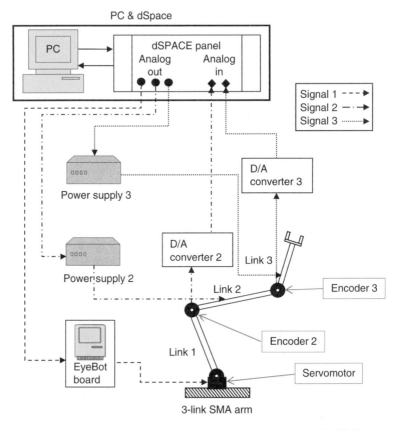

Figure 4.6 Experimental setup for position control of the 3DOF SMA arm

Table 4.2 Data for the SMA wire and actuator geometry

Data type	Value
r_{p2}	0.01075 (m)
r_{p3}	0.00825 (m)
r_2	0.01×10^{-2} (m)
r_3	0.0075×10^{-2} (m)
A_2	0.000314×10^{-4} (m^2)
A_3	0.000177×10^{-4} (m^2)
σ_{c_2}	0.0006751×10^{-6} (m^3)
σ_{c_3}	0.0002921×10^{-6} (m^3)
ξ_{02} & ξ_{03}	0.04
l_{02} & l_{03}	1 (m)

Table 4.3 Data for the constitutive law and phase transformation parameters

Data type	Value
M_s	52 (°C)
A_s	68 (°C)
A_f	78 (°C)
M_f	42 (°C)
C_A & C_M	10.3×10^6 (Pa/°C)
a_A & a_M	0.314 (/°C)
b_A & b_M	-0.0304×10^{-6} (/Pa)
Ω_2 & Ω_3	-2.06×10^9 (Pa)
θ_T	-0.055×10^6 (Pa/°C)
E_A	75×10^9 (Pa)
E_M	28×10^9 (Pa)

Table 4.4 Data for the heat transfer and electrical parameters

Data type	Value
c_{p2} & c_{p3}	837.4 (J/(kg-K))
ρ_2 & ρ_3	6.45 (kg/m^3)
m_{w3}	0.11392×10^{-3} (kg)
m_{w2}	0.20253×10^{-3} (kg)
A_{l_2}	6.28×10^{-4} (m^2)
A_{l_3}	4.71×10^{-4} (m^2)
h_2 & h_3	85 (W/(m^2-K))
R_2	28.575 (ohm)
R_3	50.8 (ohm)
T_∞	+20 (°C)

Table 4.5 Geometric and inertia properties of the 3DOF robot

Data type	Value
m_1	30×10^{-3} (kg)
m_2	28×10^{-3} (kg)
m_3	105.65×10^{-3} (kg)
l_1	11×10^{-2} (m)
l_2	7×10^{-2} (m)
l_3	7×10^{-2} (m)
d_1	7.33×10^{-2} (m)
d_2	4.75×10^{-2} (m)
d_3	5.51×10^{-2} (m)
I_1	201.33×10^{-6} (kg-m^2)
I_2	131.36×10^{-5} (kg-m^2)
I_3	177.24×10^{-4} (kg-m^2)

In this case, the surface for each actuator may be defined as

$$s_i = \lambda_{P_i} \tilde{\theta} + \lambda_{V_i} \dot{\tilde{\theta}} + \lambda_{I_i} \int_0^t \tilde{\theta} dt, \ i = 2,3 \tag{4.14}$$

where the strictly positive parameters λ_{P_i}, λ_{V_i}, and λ_{I_i} can be thought of position, velocity, and integral gains, respectively. It should be emphasized that the second and third links are actuated by the SMA wire ($i = 2$, 3) and the first link is driven by a servomotor.

Since the control law is not model based, the range of acceptable values for control parameters (boundary layer thicknesses, ϕ_i, and surfaces parameters, λ_{P_i}, λ_{V_i}, and λ_{I_i}) must be experimentally determined. Figure 4.7 shows the controller performance for boundary layer thicknesses of 0–15° for both ϕ_2 and ϕ_3. Clearly, the boundary layer θ_2 must be larger to avoid overshoot ($\phi_2 = 15$), while θ_3 should have a smaller boundary layer for a fast response ($\phi_3 = 8$). Variations in position gains show that larger gains are necessary for joint 2 ($6 \leq \lambda_{P_2} \leq 12$) when compared to joint 3 ($1 \leq \lambda_{P_3} \leq 5$) in order to strike a balance between overshoot and steady-state error, as shown in Figure 4.8. However, smaller velocity gains are required for joint 2 ($1 \leq \lambda_{V_2} \leq 2$) compared to joint 3 ($1 \leq \lambda_{V_3} \leq 4$), as shown in Figure 4.9. The integral gain λ_I must be relatively small and can only reduce steady-state error, as shown in Figure 4.10. The gains resulting in the smallest steady-state errors seem to be $0.002 \leq \lambda_{I_2} \leq 0.003$ for

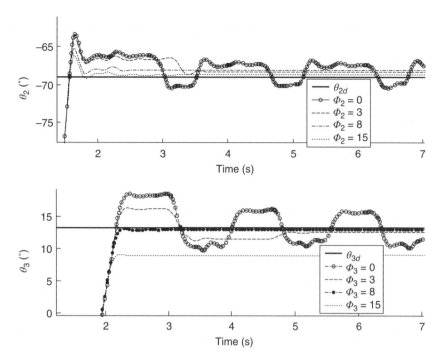

Figure 4.7 Effect of boundary layer thickness on the controller performance

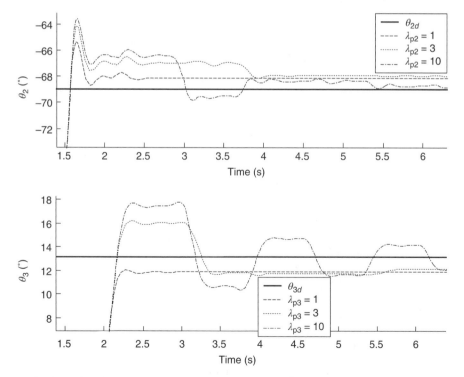

Figure 4.8 Effect of surface position gain on the controller performance

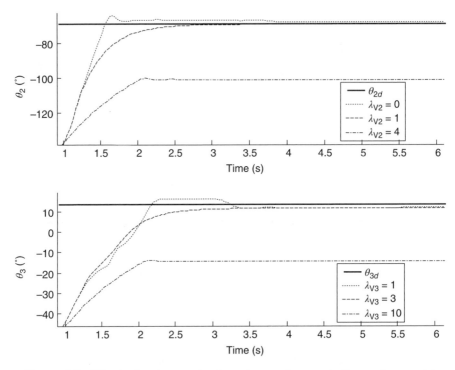

Figure 4.9 Effect of surface velocity gains on the controller performance

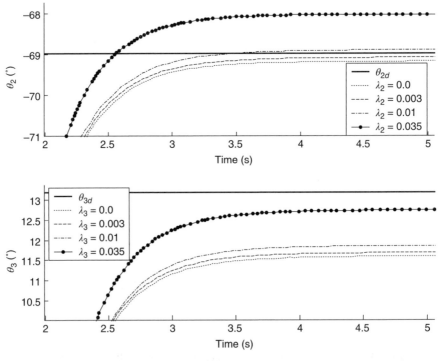

Figure 4.10 Effect of surface integral gains on the controller performance

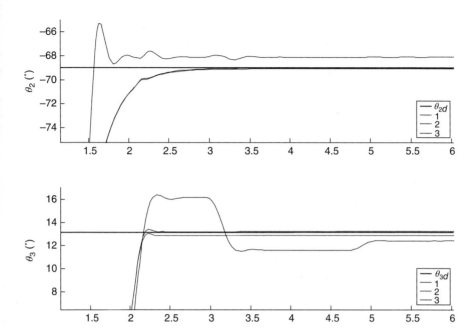

Figure 4.11 Comparison of the sliding mode controllers with 1-position, 2-position and velocity, 3-position, velocity and integral error feedbacks

joint 2 and $0.030 \leq \lambda_{I_3} \leq 0.035$ for joint 3. The evolutionary improvement in performance with only position feedback, position plus velocity feedback, and position plus velocity and integral are shown in Figure 4.11. It is clear that the new surface with integral gain has a superior performance.

4.3 Model-Based Sliding Control of SMA Actuators

Control methods that do not rely on model are easy to implement, robust, and require little calculations. However, there are several shortcomings associated with these methods. These methods lack rigorous stability proof. Furthermore, they are normally application specific and lack good performance in tracking trajectories. Hence, in this section, we present an asymptotically stable sliding mode control law for generic mechanical systems actuated by SMA. For simplicity, we assume the dynamics of these systems can be represented by a series of ordinary differential equations. We will present the model which includes the system's equations of motion, SMA constitutive law, heat convection, and phase transformation equations. We then apply the control law to the

3DOF robot arm introduced in the previous section to demonstrate its performance for both set point and trajectory tracking.

Equations of Motion

The equations of motion of an nDOF mechanical system can be expressed in terms of its $n \times 1$ position, velocity, and acceleration vectors, $q, \dot{q},$ and $\ddot{q}$, as

$$M(q)\ddot{q} = f_q(q,\dot{q}) + \tau(\sigma) \qquad (4.15)$$

where M is the mass matrix, f_q is the $n \times 1$ force vector, τ is the $n \times 1$ actuator forces, and σ is the $n \times 1$ SMA stress vector.

The Constitutive Law

The SMA constitutive model, in a rate form, relates the rate of change of stress $\dot{\sigma}$, the strain rate $\dot{\varepsilon}$, the martensite fraction rate $\dot{\xi}$, and the rate of change of temperature $\dot{T}$:

$$\dot{\sigma} = E(\xi)\dot{\varepsilon} + \Theta_T \dot{T} + \Omega \dot{\xi} \qquad (4.16)$$

where E, Θ_T, and Ω are the $n \times n$ diagonal nonsingular matrices of Young's moduli, thermal expansion factors, and phase transformation contribution factors, respectively. SMA actuators' strain is a function of system position vector as $\varepsilon = \alpha q$, where α is a diagonal positive definite matrix, which depends on the SMA actuator geometry.

Convection Heat Transfer

The SMA wire convection heat transfer model determines the rate of heating and cooling due to changes in applied electrical voltage. The rate of heat transfer due to natural convection may be presented as

$$T = f_T(T)(T - T_\infty) + g_T u \qquad (4.17)$$

where T_∞ is the $n \times 1$ vector containing room temperature and f_T and g_T are $n \times n$ diagonal nonsingular matrices, which are functions of geometric, thermal, and electrical properties of SMA. The input vector includes the squared of the electrical voltages:

$$u = \begin{bmatrix} V_1^2 \dots V_n^2 \end{bmatrix}$$

SMA Phase Transformation

The reverse transformation from martensite to austenite during heating and forward transformation from austenite to martensite during cooling are modeled macroscopically as a function of SMA temperature and stress. The stress-induced martensite fraction in reverse transformation ($\dot{\xi}_i < 0$ and $\dot{T}_i > 0$) for each actuator i is given as

$$\xi_i = f_{r\xi_i}(\sigma, T_i) \text{ if } A_{si}(\sigma_i) \le T_i \le A_{fi}(\sigma_i) \tag{4.18}$$

where A_s and A_f are austenite phase start and final temperatures, respectively. The rate of change of martensite fraction is determined using the time derivative of Equation 4.18:

$$\dot{\xi} = h_{r\sigma}(\sigma, T)\dot{\sigma} + h_{rT}(\sigma, T)\dot{T} \tag{4.19}$$

where $h_{r\sigma}$ & h_{rT} are $n \times n$ diagonal matrices of partial derivatives of $f_{r\xi_i}$ with respect to σ_i and T_i, respectively.

Similarly, the forward transformation equation ($\dot{\xi}_i > 0$ & $\dot{T}_i < 0$) for each SMA actuator i may be written as

$$\xi_i + f_{f\xi_i}(\sigma_i, T_i) \text{ if } M_{fi}(\sigma_i) \le T_i \le M_{si}(\sigma_i) \tag{4.20}$$

where M_s and M_f, are martensite phase start and final temperatures, respectively. Taking the time derivative of Equation 4.20, the rate of change of stress-induced martensite fraction can be determined as

$$\dot{\xi} = h_{f\sigma}(\sigma, T)\dot{\sigma} + h_{fT}(\sigma, T)\dot{T} \tag{4.21}$$

where $h_{f\sigma}$ & h_{fT} are $n \times n$ diagonal matrices of partial derivatives of $f_{f\xi_i}$ with respect to σ_i and T_i, respectively. Based on these direct algebraic relationships, martensite fraction ξ is not required to be part of the state vector.

Sliding Mode Control Law

The objective of the control law is for the system to track a desired position. Since the stress rate $\dot{\sigma}$ is proportional to the control u, we take the time derivative of Equation 4.15 to directly relate the system motion to input u:

$$M(q)\dddot{q} = \dot{f}_q(q, \dot{q}, \ddot{q}) - \dot{M}(q, \dot{q})\ddot{q} + \tau_\sigma \dot{\sigma} \tag{4.22}$$

where τ_σ is a diagonal matrix of partial derivatives of τ with respect to σ. Next, we define asymptotically stable second-order sliding surfaces of position tracking errors:

$$S = \ddot{\tilde{q}} + 2\lambda\dot{\tilde{q}} + \lambda^2\tilde{q} \qquad (4.23)$$

where $\tilde{q} = q - q_d$, q_d is the desired position vector, and λ is a diagonal positive definite matrix. Following the standard procedure for sliding mode control approach, let $\dot{s} = 0$:

$$\dot{S} = \dot{q} - \dot{s}_r = 0, \quad \dot{s}_r = \ddot{q}_d - 2\lambda\ddot{\tilde{q}} - \lambda^2\dot{\tilde{q}} \qquad (4.24)$$

Substituting from Equations 4.16, 4.17, and 4.22 into Equation 4.24 and adding the robustness term, we get

$$u = \hat{B}^{-1}\left[-\hat{f} + \hat{M}\dot{s}_r - k\,\mathrm{sgn}(S)\right] \qquad (4.25)$$

where " $\hat{\ }$ " denotes the nominal model, and

$$B = \tau_\sigma(I_n - \Omega h_\sigma)^{-1}(\Theta_T + \Omega h_T)g_T,$$

$$f = \dot{f}_q - \hat{M}\ddot{q} + \tau_\sigma(I_n - \Omega h_\sigma)^{-1}[E\dot{e} + |\Theta_T + \Omega h_T)f_T(T - T_\infty)],$$

$$k\,\mathrm{sgn}(S) = \left[k_1\mathrm{sgn}(S_1)\ldots k_n\mathrm{sgn}(S_n)\right]^T$$

In the above equation, I_n is the identity matrix of size n, and the parameters h_σ and h_T are defined based on the type of transformation:

$$h_\sigma(\text{or } h_T) = \begin{cases} h_{r\sigma}\left(\text{or } h_{fT}\right) \text{reverse transformation} \\ h_{f\sigma}\left(\text{or } h_{fT}\right) \text{forward transformation} \end{cases}$$

The gains of the signum functions k are calculated to ensure that the reaching condition for all surfaces ($S_i\dot{S}_i \le -\eta_i|S_i|, \eta_i > 0, i = 1\ldots n$) are satisfied. Defining the uncertainty bounds:

$$|f_i - \hat{f}_i| \le F_i, \quad i = 1\ldots n, \; j = 1\ldots n$$

$$B = (I + \delta)\hat{B}, \; |\delta_{ij}| \le \Delta_{ij}$$

where Δ and F define the bounds on uncertainties in matrix B and vector f, respectively. The gain vector is therefore given as

$$k = (I_n + \Delta)^{-1} \left[\eta + F + \Delta \left| M s_r - \hat{f} \right| \right] \tag{4.26}$$

The control law is only applied when the SMA is in the phase transformation/ actuation region. Otherwise, either the maximum voltage for heating or no voltage for cooling is applied until the actuation region is reached. In addition, the discontinuous signum functions may be replaced with continuous saturation functions of boundary layers $\phi = [\phi_1 \dots \phi_n]^T$ to avoid chattering:

$$u = \begin{cases} B^{-1} \left[-\hat{f} + M \dot{s}_r - k \tan h\left(\dfrac{s}{\phi}\right) \right] & \text{actuation region} \\[2mm] \text{Otherwise} \quad \begin{cases} u_{\max} & \text{heating} \\[1mm] 0 & \text{cooling} \end{cases} \end{cases} \tag{4.27}$$

where $k \tan h(S/\phi) = [k_1 \tan h(S_1/\phi_1) \dots k_n \tan h(S_n/\phi_n)]$.

Note that SMA actuator stresses and temperatures are difficult to measure and therefore must be estimated in most cases. The reader is referred to works in Refs. [19, 26] for fairly successful estimation methods, based on an extended Kalman filter (EKF), using only position measurements.

Example

We applied the model-based control law to the 3DOF robotic introduced earlier in this chapter. The system is comprised of eight states in this case:

$$x = \begin{bmatrix} \theta_2 & \theta_3 & \dot{\theta}_2 & \dot{\theta}_3 & \sigma_2 & \sigma_3 & T_2 & T_3 \end{bmatrix}^T$$

Of all the eight states, only the two joint angles are measured using position encoders, while the rest are estimated using EKF. The control parameters selected for both problems are $\lambda_2 = \lambda_3 = 4$, $\eta_2 = 240$, $\eta_3 = 300$, and $\phi_2 = \phi_3 = 4°$, and the maximum applied voltage is 20 V.

We have defined a two-stage problem. In the first stage, the robot joints 2 and 3 are required to go from their initial angles of $-146°$ and $-60°$ to desired positions $-68°$ and $13°$, respectively, which would require electrical heating.

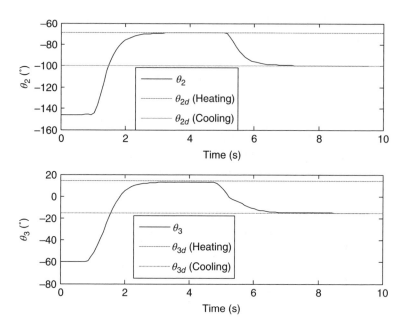

Figure 4.12   Joint angle time histories for two set points requiring heating and cooling, respectively

In the second stage, the joints must rotate back to some intermediate positions at −100° and −15°, respectively, which would require natural cooling. Figure 4.12 shows the performance of the controller in rotating the joints as commanded. Figure 4.13 shows how the SMA wire temperatures reach and settle in the heating and cooling actuation regions. Figure 4.14 shows the control input voltage variations.

For trajectory tracking problem, we start from the same initial conditions and require the joints to reach the same final angles as the two stages of the set position problem. However, the joints must reach their desired angles in 15 s and follow fifth-order polynomial trajectories during each stage. The coefficients of the polynomials are defined based on starting and ending at rest. The control parameters used for trajectory tracking are $\lambda_2 = \lambda_3 = 1$, $\eta_2 = \eta_3 = 50$, and $\phi_2 = \phi_3 = 5°$, and the maximum voltage in this case is 10 V. Figure 4.15 shows the excellent performance of the controller in tracking the trajectory. Figure 4.16 reveals that the tracking error never exceeds half a degree. Figure 4.17 shows the SMA wire temperatures reaching the actuation regions and following the corresponding trajectories.

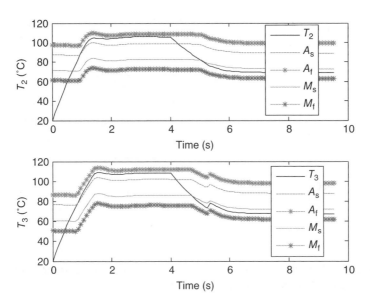

Figure 4.13 Temperature changes for two set points requiring heating and cooling, respectively

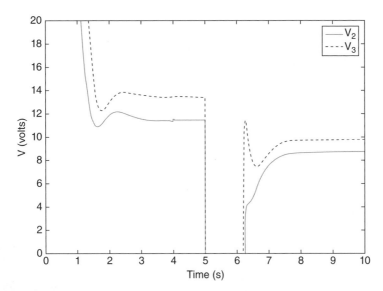

Figure 4.14 Time histories input voltages during trajectory tracking control

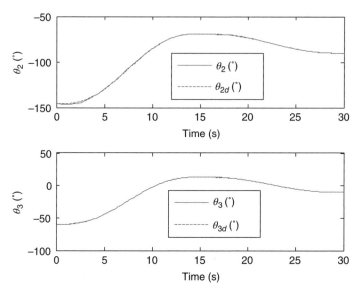

Figure 4.15 Joint angle time history for forward and reverse trajectory tracking. The subscript "d" denotes desired trajectories

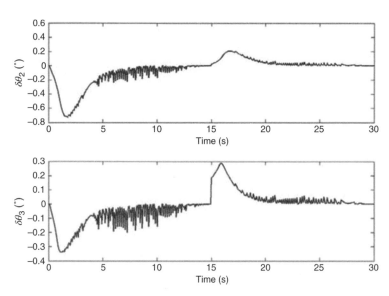

Figure 4.16 Error in joint angle time history for forward and reverse trajectory tracking

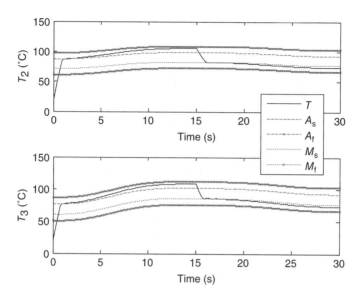

Figure 4.17 Time histories of SMA wire temperatures and phase transformation boundaries during trajectory tracking control

4.4 Model-Based Backstepping Control of SMA Actuators

Another model-based control method that has been applied for SMA actuators is backstepping. In this stabilizing technique, the control input trickles down through a series of integrators to the level of fundamental subsystem. To this end, a stabilizing control law for the fundamental subsystem is designed. Consequently, backstepping through the integrators is performed to obtain the true control law [27].

Details on the methodology can be found in [27]. Briefly, a class of single-input control systems can be modeled as

$$\dot{\eta} = f(\eta) + g(\eta)\zeta$$

$$\dot{\zeta} = u \tag{4.28}$$

where $\eta \in R^n$, $\zeta(t) \in R$, and $f(0) = 0$. The objective in designing the controller is to stabilize the origin $(\eta, \zeta) = (0,0)$. We assume that it is possible to design a feedback controller for the $\dot{\eta}$ subsystem by treating ζ as the input. By choosing

$\zeta = \phi(\eta)$ with $\phi(0) = 0$, the origin of the system is stabilized. The origin is asymptotically stable if [27]

$$u(\eta, \zeta) = -\frac{\partial V}{\partial h}g(\eta) - k[\zeta - \phi(\eta)] + \frac{\partial \phi}{\partial \eta}[f(\eta) + g(\eta)\zeta] \qquad (4.29)$$

where $V(\eta)$ is a Lyapunov function, which proves asymptotical stability for the closed-loop subsystem $\dot{\eta}$ of Equation 4.28.

The benefit of backstepping control is that it can globally asymptotically stabilize SMA actuators such as the one-DOF rotary actuator that was introduced in Chapter 1 (Figures 1.18 and 4.1). In this approach, the stress of SMA wire is assumed to be the control input of the system. It is possible to control the stress (desired stress) to asymptotically stabilize the desired angular position of the arm. The applied voltage to the SMA wire however is the actual control input. Using the constitutive, phase transformation, and heat transfer models, this voltage is calculated based on the desired stress to globally asymptotically stabilize the system [16].

The SMA wire's stress is the input for the dynamic subsystem ($\sigma = \zeta$) of

$$I\ddot{\theta} + c\dot{\theta} + \left[\tau_g(\theta) + \tau_s(\theta)\right] = \tau_w(\sigma) \qquad (4.30)$$

For simplicity, Equation 4.30 can be written as

$$I\ddot{\theta} + h(\theta, \dot{\theta}) = \alpha\sigma \qquad (4.31)$$

where $h(\theta, \dot{\theta})$ includes viscous damping and the spring and gravitational torques and $\alpha = 2r_p A_w$ is a constant relating the stress and the torque created by the SMA wire. The applied voltage as the actual control input of the system appears in the heat transfer model. The stress is indirectly related to this input through the thermomechanical model of the wire, which can be simplified as

$$\sigma = E(\xi)(\varepsilon - \varepsilon_L \xi)$$
$$E = \xi E_M + (1 - \xi)E_A \qquad (4.32)$$

As shown in Chapter 1, change of the strain of the SMA wire is associated with the rotation of the arm:

$$\varepsilon = -\frac{2r_p(\theta - \theta_0)}{l_0} + \varepsilon_L$$

$$\dot{\varepsilon} = -\frac{2r_p\dot{\theta}}{l_o} \tag{4.33}$$

where r_p is pulley's radius, l_0 is the initial length of SMA wire, θ_0 is the initial position of the arm, and ε_L is the maximum (transformation/residual) stain of the wire.

The backstepping method is applied to the system in this section. The equation of motion (4.31) can be written as two first-order equations:

$$\dot{x}_1 = x_2$$

$$\dot{x}_2 = -\frac{h(x_1, x_2)}{I} + \frac{\alpha x_3}{I} \tag{4.34}$$

where $x_1 = \theta$, $x_2 = \dot{\theta}$, and $x_3 = \sigma$. Next, we stabilize the origin, assuming the stress as the control input. We define a Lyapunov function

$$V = \frac{1}{2}x_1^2 + \frac{1}{2}x_2^2$$

$$\dot{V} = x_1\dot{x}_1 + x_2\dot{x}_2$$

$$\dot{V} = x_1 x_2 - \frac{h}{I}x_2 + \frac{\alpha}{I}x_3 x_2 \tag{4.35}$$

by designing

$$x_3 = \frac{I}{\alpha}\left(-x_1 + \frac{h}{I} - Kx_2\right) = \Phi(x_1 + x_2)$$

$$\dot{V} = -Kx_2^2 \le 0 \tag{4.36}$$

where $K > 0$. $x_3 = \sigma = \Phi(x_1, x_2)$ makes the origin $(\theta = 0,\ \dot{\theta} = 0)$ stable. Let's define a new variable to measure the tracking of the stress that produces the desired stabilization behavior:

$$z_1 = x_3 - \Phi(x_1, x_2) \tag{4.37}$$

The applied voltage to the SMA wire should satisfy $z_1 \to 0$. Taking the time derivative shows that $\dot{z}_1 = \dot{x}_3 - \dot{\Phi}(x_1, x_2)$. In order to find $\dot{x}_3 = \dot{\sigma}$, we need to rewrite the constitutive equation

$$\dot{x}_3 = \dot{x}_4 \left[(E_M - E_A)\varepsilon_L(1 - 2x_4) - \frac{2r_p}{l_0}(x_1 - x_{10})(E_M - E_A) - E_A\varepsilon_L \right]$$
$$- \frac{2r_p}{l_o} x_2[(E_M - E_A)x_4 + E_A] \tag{4.38}$$

where $x_{10} = \theta_0$ initial angular position and $x_4 = \xi$. Let's define a new variable

$$\Gamma = (E_M - E_A)\varepsilon_L(1 - 2x_4) - \frac{2r_p}{l_0}(x_1 - x_{10})(E_M - E_A) - E_A\varepsilon_L$$

$$\dot{x}_3 = \dot{x}_4 \Gamma - \frac{2r_p}{l_o} x_2[(E_M - E_A)x_4 + E_A] \tag{4.39}$$

The proper choice of martensite fraction therefore is

$$\dot{x}_4 = \frac{1}{\Gamma}\left[u_1 + \dot{\Phi}(x_1, x_2)\right] + \frac{2r_p}{l_o} x_2[(E_M - E_A)x_4 + E_A] \tag{4.40}$$

that will guarantee

$$\dot{z}_1 = u_1 \tag{4.41}$$

When $z_1 = 0$, the system of Equation 4.34 is stabilized. The next step in the design process is to choose the control law u_1 to drive z_1 to zero while keeping η bounded. To guarantee the stability to be independent from the path that z_1 assumes in approaching zero, we define an augmented Lyapunov function:

$$V_a = \frac{1}{2}x_1^2 + \frac{1}{2}x_2^2 + \frac{1}{2}z_1^2 \tag{4.42}$$

$$\dot{V}_a = x_1\dot{x}_1 + x_2\dot{x}_2 + z_1\dot{z}_1$$

Recall that

$$\dot{x}_2 = -\frac{h}{I} + \frac{\alpha x_3}{I} = -\frac{h}{I} + \frac{\alpha(\Phi + z_1)}{I}$$

$$\Phi(x_1, x_2) = \frac{I}{\alpha}\left(-x_1 + \frac{h}{I} - Kx_2\right)$$

$$\dot{V}_a = x_1 x_2 - \frac{h}{I} x_2 - x_1 x_2 + \frac{h}{I} x_2 - K x_2^2 + \frac{\alpha}{I} z_1 x_2 + z_1 u_1$$

$$\dot{V}_a = -K x_2^2 + \frac{\alpha}{I} z_1 x_2 + z_1 u_1$$

V_a is a Lyapunov function if

$$u_1 = -\frac{\alpha}{I} x_2 - k_1 z_1 \tag{4.43}$$

$$\dot{V}_a = -K x_2^2 - k_1 z_1^2 \leq 0$$

provided that $k_1 > 0$. From Equation 4.40, we can find

$$\dot{x}_2 = \frac{1}{\Gamma} \left[-\frac{\alpha}{I} x_2 - k_1 z_1 + \dot{\Phi}(x_1, x_2) \right] + \frac{2 r_p}{l_o} x_2 [(E_M - E_A) x_4 + E_A] \tag{4.44}$$

is the proper choice to guarantee stability of the SMA actuator. The martensitic phase transformation kinetic, as was explained in Chapter 1, can be captured by the two phenomenological equations. For the forward transformation

$$\xi = \frac{1 - \xi_0}{2} \cos \left[a_M \left(T - M_f - \frac{\sigma}{C_M} \right) \right] + \frac{1 + \xi_0}{2} \tag{4.45}$$

and for the reverse transformation

$$\xi = \frac{\xi_0}{2} \left[\cos \left[a_A \left(T - A_s - \frac{\sigma}{C_A} \right) \right] + 1 \right] \tag{4.46}$$

For the forward transformation, therefore, we can find the desired temperature $x_5 = T$ of the wire using Equation 4.45:

$$x_5 = \frac{1}{a_A} \cos^{-1} \frac{2 x_4 - x_{40}}{x_{40}} + A_s + \frac{x_3}{C_A}$$

$$\dot{x}_5 = \frac{1}{a_A} \frac{-2 \dot{x}_4}{\sqrt{1 - \left[\frac{2 x_4 - x_{40}}{x_{40}} \right]^2}} + \frac{\dot{x}_3}{C_A} \tag{4.47}$$

The actual input to the SMA wire is the applied voltage, which can be calculated through the heat transfer model

$$U(x_1, x_2, x_3, \dot{x}_3, x_4) = \frac{V^2}{R} = mc_p \dot{x}_5 + h_c A_c (x_5 - x_{50}) \qquad (4.48)$$

where x_{50} is the ambient temperature. Both x_5 and $\dot{x}_5$ are known from Equation 4.47. The control input therefore can be found

$$V = \sqrt{UR} \qquad (4.49)$$

Figure 4.18 shows results of the simulation of the SMA actuated arm using the backstepping control in stabilizing the origin.

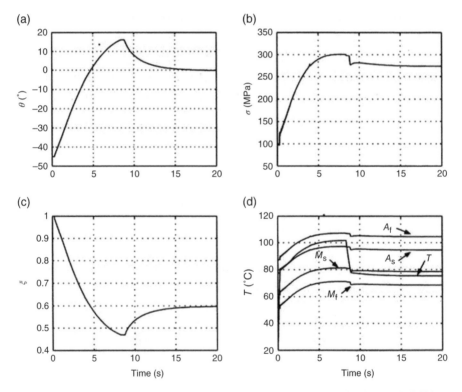

Figure 4.18 Result of simulation with a backstepping controller in stabilizing the arm. Reproduced with permission from Ref. [16], SAGE Publications. (a) Position, (b) stress, (c) martensite fraction, and (d) temperature of the SMA wire and the four transformation temperatures

4.5 Summary

SMA actuators offer flexibility and simplicity in the resulting smart structures and devices. This chapter provided a brief overview of the control challenges and existing control methodologies. Both nonmodel-based (variable structure control) and model-based (sliding model and backstepping) robust control algorithms are presented and discussed in the context of rotary SMA actuators. These actuators are chosen to provide continuity throughout the book. Additionally, the SMA elements in these types of actuators undergo complex thermomechanical loadings. This in turn adds another level of complexity in the control of these systems, as demonstrated in this chapter.

References

[1] Arai K, Aramaki S, Yanagisawa K. Continuous system modeling of shape memory alloy (SMA) for control analysis. Fifth International Symposium on Micro Machine and Human Science, Nagoya, Japan. IEEE; 1994, p. 97.

[2] Gorbet RB, Wang DWL. General stability criteria for a shape memory alloy position control system.IEEE International Conference on Robotics and Automation, Nagoya, Japan: IEEE; 1995. p. 2313–9.

[3] Shameli E, Alasty A, Salaarieh H. Stability analysis and nonlinear control of a miniature shape memory alloy actuator for precise applications. Mechatronics 2005;15:471–86.

[4] Gédouin P-A, Delaleau E, Bourgeot J-M, Join C, Arbab Chirani S, Calloch S. Experimental comparison of classical pid and model-free control: position control of a shape memory alloy active spring. Control Engineering Practice 2011;19:433–41.

[5] Esfahani ET, Elahinia MH. Developing an adaptive controller for a shape memory alloy walking assistive device. Journal of Vibration and Control 2010;16:1897–914.

[6] Gharaybeh MA, Burdea GC. Investigation of a shape memory alloy actuator for dextrous force-feedback masters. Advanced Robotics 1994;9:317–29.

[7] Honma D, Miwa Y, Iguchi N. Micro robots and micro mechanisms using shape memory alloy. The Third Toyota Conference, Integrated Micro Motion Systems, Micro-Machining, Control and Applications Nissan, Aichi, Japan; 1984.

[8] Choi S-B, Cheong C-C. Vibration control of a flexible beam using shape memory alloy actuators. Journal of Guidance, Control, and Dynamics 1996;19:1178–80.

[9] Dickinson CA, Wen JT. Feedback control using shape memory alloy actuators. Journal of Intelligent Material Systems and Structures 1998;9:242–50.

[10] Kilicarslan A, Grigoriadis KM, Song G. Nonlinear Control of a Shape Memory Alloy Actuator via Mu-Synthesis. Proceedings of Earth and Space 2010-12th International Conference on Engineering, Science, Construction, and Operations in Challenging Environments, Honolulu, HI. ASCE; 2010, 1662–1675.

[11] Grant D, Hayward V. Variable structure control of shape memory alloy actuators. IEEE Control Systems Magazine 1997;**17**:80–8.

[12] Nakazato T, Kato Y, Masuda T. Control of push-pull-type shape memory alloy actuators by fuzzy reasoning. Transactions of Japan Society of Mechanical Engineers, Part C 1993;**59**:226–32.

[13] Song G, Chaudhry V, Batur C. Precision tracking control of shape memory alloy actuators using neural networks and a sliding-mode based robust controller. Smart Materials and Structures 2003;**12**:223.

[14] Van der Wijst MWM, Schreurs PJG, Veldpaus FE. Application of computed phase transformation power to control shape memory alloy actuators. Smart Materials and Structures 1997;**6**:190.

[15] Reynaerts D, Brussel HV. Design aspects of shape memory actuators. Mechatronics 1998;**8**:635–56.

[16] Elahinia M, Koo J, Ahmadian M, Woolsey C. Backstepping control of a shape memory alloy actuated robotic arm. Journal of Vibration and Control 2005;**11**:407–29.

[17] Elahinia MH, Seigler TM, Leo DJ, Ahmadian M. Nonlinear stress-based control of a rotary SMA-actuated manipulator. Journal of Intelligent Material Systems and Structures 2004;**15**:495–508.

[18] Elahinia MH, Ashrafiuon H, Ahmadian M, Tan H. A temperature-based controller for a shape memory alloy actuator. Journal of Vibration and Acoustics 2005;**127**:285–91.

[19] Ashrafiuon H, Jala VR. Sliding mode control of mechanical systems actuated by shape memory alloy. Journal of Dynamic Systems, Measurement, and Control 2009;**131**:011010.

[20] Elahinia MH, Esfahani ET, Wang S. Control of SMA Systems: Review of the State of the Art. In: Chen RH, editor. Shape Memory Alloys: Manufacture, Properties and Applications. New York: Nova Science Publishers, Inc.; 2010. p. 49–68

[21] Vadim IU. Survey paper variable structure systems with sliding modes. IEEE Transactions on Automatic Control 1977;**22**:212–222.

[22] Hung JY, Gao W, Hung JC. Variable structure control: a survey. IEEE Transactions on Industrial Electronics 1993;**40**:2–22.

[23] Slotine J-JE, Li W. Applied Nonlinear Control. Englewood Cliffs, NJ: Prentice-Hall; 1991.

[24] Elahinia M, Ashrafiuon H. Nonlinear control of a shape memory alloy actuated manipulator. Journal of Vibration and Acoustics 2002;**124**:566–75.

[25] Ashrafiuon H, Eshraghi M, Elahinia M. Position control of a three-link shape memory alloy actuated robot. Journal of Intelligent Material Systems and Structures 2006;**17**:381–92.

[26] Elahinia M, Ahmadian M, Ashrafiuon H. Design of a Kalman filter for rotary shape memory alloy actuators. Smart Materials and Structures 2004;**13**:691.

[27] Khalil HK, Grizzle JW. Nonlinear Systems. Upper Saddle River, NJ: Prentice hall; 2002.

5

Fatigue of Shape Memory Alloys

Mohammad J. Mahtabi, Nima Shamsaei
and Mohammad H. Elahinia

Fatigue is generally defined as the gradual deterioration of materials under cyclic loading. This type of failure is very important because of its unexpected nature. It can cause serious damage to mechanical and biomedical systems without any warning. In the literature, fatigue has been found to be responsible for 50–90% of all mechanical failures [1].

As discussed in the previous chapters, shape memory alloys have widespread applications in various industries such as automotive, aerospace, civil, and bioengineering. The realistic loading condition in many of these applications is cyclic; therefore, fatigue is often the main failure mode for such components and structures. Two prominent superelastic NiTi medical devices are the stents to expand the blood vessels and the dental files for root canal treatments. Expansion and contraction of the blood vessels and the torsional loadings in dental files constitute as cyclic loading in these SMA elements. Some early studies have considered the fatigue of SMA actuators and confirmed a low-cycle fatigue behavior in these components [2].

The first article addressing the fatigue behavior of an SMA was published in 1958 for a single-crystal Cu–Al–Ni in plane bending [3]. By discovering

Shape Memory Alloy Actuators: Design, Fabrication, and Experimental Evaluation, First Edition. Mohammad H. Elahinia.
© 2016 John Wiley & Sons, Ltd. Published 2016 by John Wiley & Sons, Ltd.

the more interesting properties of the SMAs, over the years, applications of these materials have been widely increased. Accordingly, more comprehensive studies on the fatigue of these alloys have been conducted to better understand their behavior under more realistic loading conditions [4–10].

Compared to other metals, SMAs can tolerate higher strain amplitudes prior to failure. While NiTi alloys may endure strain amplitudes ranging from 4 to 12%, for most other metals, the total tolerable strain before failure is usually around 1% for the same number of load cycles [11]. A single crystal of Cu–Al–Ni SMA under a strain amplitude of 2% lasts about 53 000 cycles, while most metals tolerate 50 cycles or less under such a strain level [3].

Fatigue behavior of SMAs is more complex compared to other materials. The complexity arises from the drastic effects of shape memory properties and the phase transformation phenomenon on the mechanical behavior of the material [12]. Any changes in the shape memory properties resulting from material compositions (e.g., amount of nickel or titanium in a NiTi alloy), heat treatment processes, or test temperature may significantly affect the fatigue behavior of these alloys. Therefore, not a general fatigue life prediction method exists for SMAs.

To protect against fatigue, the miniature components made from SMAs require the control of the crack initiation rather than the crack growth. On the other hand, difficulties concerned with experimental work and measurements have been an obstacle for monitoring the crack growth behavior. Currently, the interest to control the crack initiation seems reasonable since the failure in these materials takes place shortly after crack nucleation. As an example, the time duration for a NiTi stent (with ~0.2 mm width) between crack nucleation and the final fracture is no longer than a few days. This is in contrast with the cases of crack propagation analysis where the geometry of the component should be large enough to retain the functionality of the components for a sufficient period of time in the presence of cracks [13]. An example of the force-controlled loading situation of NiTi is in orthodontics, where the alloy is used for *in vivo* dental stabilizing devices. These devices have dimensions of several centimeters, and the fatigue crack growth could be a more appropriate approach for durability analysis of these components.

In this chapter, a review on the fatigue behavior of metals and its various aspects including fatigue design approaches and fatigue testing methods are presented. This is followed by an attempt to characterize the fatigue behavior of SMAs with an emphasis on NiTi fatigue. To this end, major fatigue models for SMAs are reviewed. Finally, a few examples of fatigue analysis of these alloys are presented and discussed.

5.1 Fatigue of Metals

Many components and structures fail under repeated or cyclic stresses. Careful analyses have revealed that the actual maximum stress in a failed component is often significantly lower than the ultimate strength, sometimes even lower than the yield strength of the material. These unexpected failures resulted from damage accumulation and gradual deterioration of the material under repeated loading over a period of time. This type of failure that occurs under repeated loads is called *fatigue failure*. A comprehensive definition of fatigue failure is provided by the ASTM standard [14] as

> The process of **progressive localized permanent** structural change occurring in a material subjected to conditions which produce **fluctuating** stresses and strains at some point or points and which may culminate in **cracks** or complete **fracture** after a sufficient number of fluctuations.

The fatigue failure typically consists of three stages: the initiation of microcracks, the propagation of cracks (or so-called crack growth stage), and the final sudden fracture. During the first stage of the fatigue failure, cracks are not normally visible by the naked eyes. Macrocracks then form from these microcracks during the crack growth state. In each cycle of loading, these cracked surfaces open and close and rub together until the final stress cycle when the cracks reach a size that the remaining cross section cannot tolerate the loads, and therefore, final fracture occurs. Depending on the material behavior and microstructural mechanism, a final fracture might be brittle, ductile, or a combination of both.

To perform fatigue characterization, a cyclic repeated load is typically applied to specimens up to the failure of the specimen. This load can be uniaxial, torsional, or multiaxial. A schematic of cyclic constant amplitude loading and related parameters is presented in Figure 5.1. The applied stresses/strains may have a zero mean stress/strain, and the cyclic load alternates between equal positive and negative magnitudes (i.e., fully reversed condition) or can have a nonzero mean stress/strain, and the loading value changes from a maximum positive or negative value to a minimum load.

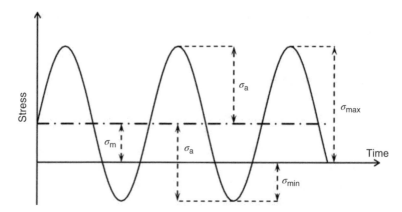

Figure 5.1 Schematic constant amplitude cyclic loading diagram

The following parameters can be defined from Figure 5.1:

σ_{max}: maximum stress

σ_{min}: minimum stress

$\sigma_m = \dfrac{\sigma_{max} + \sigma_{min}}{2}$: mean stress, which is the average of σ_{max} and σ_{min}

$\Delta\sigma = \sigma_{max} - \sigma_{min}$: stress range, which is the difference of the σ_{max} and σ_{min}

$R = \dfrac{\sigma_{min}}{\sigma_{max}}$: stress ratio, which is the ratio of σ_{min} to σ_{max}

$\sigma_a = \sigma_{max} - \sigma_m$: alternating stress or stress amplitude, which is the difference of σ_{max} and σ_m

Similar parameters can be defined by replacing the stress with strain for the strain loading diagram.

In order to perform a reliable fatigue study, a series of experiments should be conducted on a number of similar specimens at various load levels. The sustained stress or strain is then plotted against the number of cycles to failure in semilog or log–log coordinate system. According to ASTM, the following parameters are defined for a fatigue curve (e.g., S–N or ε–N curve). The fatigue life, N_f, is the number of cycles of a specified character (e.g., stress or strain) that a given specimen sustains before failure. Fatigue strength, σ_f, is a hypothetical value of a stress at failure for exactly N_f cycles. In certain materials as the stress level is reduced, a value may be reached at which the failure does not occur, regardless of the number of applied cycles. This stress level is called the fatigue limit or the endurance limit for the material under study. According to

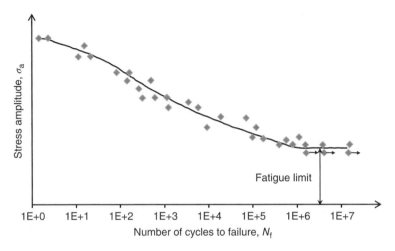

Figure 5.2 Typical stress-life fatigue plot for a material with endurance limit. Arrows show specimens for which fatigue failure did not occur

ASTM Standard E1823, the endurance limit, σ_e, is the limiting value of the median fatigue strength as the fatigue life becomes very large. Typical fatigue curve with endurance limit is shown in Figure 5.2. In this figure, σ_a is the stress amplitude and N_f is the number of cycles to failure. Details on different approaches for fatigue life predictions of materials are presented in following sections.

5.1.1 Fatigue Approaches

Generally, there are four approaches for fatigue analysis and life prediction. These approaches are the nominal stress-life ($S–N$) method, local strain-life ($\varepsilon–N$) method, fatigue crack growth ($da/dN – \Delta K$) method, and the two-stage method which combines the stress (or local strain) life and the fatigue crack growth methods for analysis and design [1].

The nominal stress-life method, which is the traditional method for the fatigue analysis and design, utilizes nominal stresses and relates them to local fatigue strengths. There are several mathematical models for the representation of stress-life fatigue curves of metals. These models usually are presented based on median fatigue life of multiple samples. One of the most commonly used models for the stress-life approach is Basquin's model [15], presented as

$$\sigma_a = \sigma'_f \left(2N_f\right)^b \tag{5.1}$$

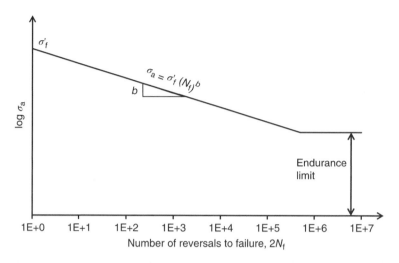

Figure 5.3   Nominal stress-life method presented by Basquin's equation

where σ_a is the applied alternating stress (i.e., stress amplitude) and σ'_f and b are material constants. The fatigue strength coefficient, σ'_f, is the stress amplitude at which the failure occurs at one reversal, and fatigue strength exponent, b, is the slope of the logarithmic linear line (Figure 5.3). Basquin's equation represents the fatigue behavior up to the endurance limit.

In the local strain method, the tests are conducted in the strain-controlled (or displacement-controlled) condition, and the local strain at a notch is correlated to the fatigue life. This method is very popular because of the fact that strains can be directly measured using strain gauges or extensometers. The strain-life method is very practical for fatigue life evaluation of the notched members. In this method, failure criterion may be defined as final fracture, a small detectable crack or a certain percentage reduction in the load amplitude.

The total measured strain amplitude ($\Delta\varepsilon/2$) can be divided into elastic strain amplitude ($\Delta\varepsilon_e/2$) and plastic strain amplitude ($\Delta\varepsilon_p/2$), as presented by Equation 5.2:

$$\frac{\Delta\varepsilon}{2} = \varepsilon_a = \frac{\Delta\varepsilon_e}{2} + \frac{\Delta\varepsilon_p}{2} \tag{5.2}$$

In the log–log coordinate system, both elastic and plastic strains typically show linear behavior with respect to the number of cycles to failure. For elastic strain, this relationship can be formulated using Basquin's relation [15] as follows:

$$\frac{\Delta \varepsilon_e}{2} = \frac{\sigma_a}{E} = \frac{\sigma_f'}{E}(2N_f)^b \tag{5.3}$$

where E is the modulus of elasticity of material. The plastic strain amplitude, $\Delta \varepsilon_p/2$, can be related to fatigue life as follows:

$$\frac{\Delta \varepsilon_p}{2} = \varepsilon_f'(2N_f)^c \tag{5.4}$$

which is called Coffin–Manson relationship [16, 17]. Finally, the total strain-life can be presented using the following equation:

$$\varepsilon_a = \frac{\sigma_f'}{E}(2N_f)^b + \varepsilon_f'(2N_f)^c \tag{5.5}$$

In Equations 5.3, 5.4, and 5.5, ε_f', σ_f', b, and c are material constants which are experimentally determined. Constant c is called the fatigue ductility exponent and constant b is the strength exponent, and ε_f' and σ_f' are the material fatigue durability and strength coefficients, respectively.

The fatigue crack growth method uses fracture mechanics concepts and theories to calculate the number of cycles required for crack propagation between two crack lengths of interest or from a crack length to the final fracture. In the two-stage approach, the local strain-life or stress-life method is used to calculate the life to the initiation of a small crack. This life will then be added to the crack growth life to determine the total fatigue life. Among all methods for fatigue analysis, the local strain-life method is the most appropriate approach for analyzing the fatigue behavior of SMAs. This fact will be comprehensively discussed in the upcoming sections.

5.1.1.1 Mean Stress/Strain Effects

The tensile or compressive mean stress/strain may significantly affect the fatigue behavior of metallic materials. The tensile mean stress usually reduces the fatigue resistance, while the compressive mean stress may increase the fatigue strength. One of the most common methods for evaluating the mean

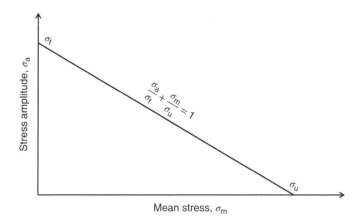

Figure 5.4 Modified Goodman model for mean stress effects on fatigue behavior

stress effects on the fatigue life is proposed by Goodman [1]. The modified Goodman equation for mean stress effects is presented as

$$\frac{\sigma_a}{\sigma_f} + \frac{\sigma_m}{\sigma_u} = 1 \tag{5.6}$$

where σ_a is the alternating stress, σ_f is the fatigue strength for zero mean stress condition ($R = -1$ or fully reversed), σ_m is the mean stress, and σ_u is the ultimate tensile stress. For the infinite life, fatigue strength, σ_f, can be replaced by the endurance limit, σ_e, in Equation 5.6. A graphical representation of the modified Goodman model is shown in Figure 5.4.

Another model for mean stress effects is proposed by Soderberg [18]. This model can be obtained by replacing the σ_u with σ_y in the Goodman model (i.e., Eq. 5.6) and is a more conservative consideration of the mean stress effects. More detailed discussions on the Goodman model and other mean stress correction methods can be found in fatigue and machine design textbooks [1, 19].

5.1.2 Fatigue Specimens

Depending on the fatigue testing method, different types of fatigue specimens should be used such as wire, flat and solid dogbone-shaped, and thin-walled tubular specimens. In addition, for some fatigue analysis methods, additional specimen preparation steps may be needed before the tests are conducted. For instance, in fatigue crack initiation behavior, the surface finish quality is very

(a) Diamond-shape   (b) Flat dogbone specimen   (c) Circular dogbone   (d) Thin-walled tubular
   components   specimen   specimen

Figure 5.5   Typical test specimens used for fatigue analysis of shape memory alloys: (a) diamond-shape components, (b) flat dogbone specimen, (c) circular dogbone specimen, and (d) thin-walled tubular specimen

important, and the specimen surface should be finely polished. Typical fatigue specimens used for SMAs are shown in Figure 5.5.

Many studies on the fatigue of SMAs to date have been employing wire specimens to conduct the simple rotary-bending tests. For other types of tests such as the compression or torsional tests, the wires are not appropriate specimens as they cannot carry compressive or torsional loads. In some studies, stent-like diamond-shaped (Figure 5.5a) components were tested under displacement-controlled fatigue loads [20, 21]. For these tests, finite element analysis is utilized to convert the applied deformation to the stress and strain terms in the critical elements of the component. Flat and circular solid dogbone-shaped specimens (Figures 5.5b and c) are more suitable candidates for compression or torsion test purposes. Tubular specimens (Figure 5.5d) are another type of specimens for fatigue analysis of SMAs, specifically for torsional studies.

5.1.3 Fatigue Testing

Fatigue testing methods can be classified based on different perspectives: uniaxial versus multiaxial fatigue, constant amplitude versus variable amplitude loading, and force-controlled versus strain- (displacement-) controlled. Standard terminology, specimen design, and procedure for performing force-controlled and strain-controlled fatigue testing of metallic materials can be found in ASTM Standards E466 [22] and E606 [23], respectively.

One of the simplest and most commonly used tests for fatigue studies of SMA wires is rotary-bending test configuration. The rotary-bending test is, in fact, a uniaxial test. Several experimental studies have used rotary-bending tests to evaluate the fatigue behavior of SMAs [24–26]. It has been proven that the results from these tests are suitable for fatigue life prediction purposes. Rotary-bending fatigue analysis has become a standard test setup for the

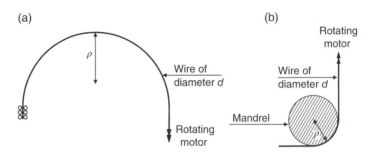

Figure 5.6 Various configurations for rotary-bending fatigue test of shape memory alloy wires: (a) non-guided with counter-rotating shafts and (b) guided fatigue apparatus

structural fatigue evaluation of shape memory wires, and an ASTM standard for rotary-bending testing of wires [27] has been recently issued. Different configurations of rotary-bending tests, used for fatigue testing of SMAs, are shown in Figure 5.6.

Since most of the experiments for fatigue analysis of SMAs have utilized wire specimens, the configuration in Figure 5.6a is the most popular method of testing. This test setup is an appropriate way to collect strain-based fatigue data for wire specimens under zero mean strain conditions. In this method, in order to apply a specific strain, the wire is bent to a certain curve with known radius of curvature (ρ). The strain amplitude is then approximated by the ratio of the wire diameter (d) to the radius of the curvature:

$$\varepsilon_a = \frac{d}{2\rho} \tag{5.7}$$

A range of various strain amplitudes can be achieved by altering the wire diameter and the radius of curvature. Alternating strains are then applied by rotating the curved wire around its longitudinal axis. The same approach has been used by a new setup by Pelton and coworkers [25] using mandrels of different sizes as the basis for the wire curvature as shown in Figure 5.6b.

5.2 Fatigue of SMAs

In addition to the structural fatigue failure mode, similar to other metallic materials, SMAs may also fail due to the loss of functionality. Their fatigue behavior is also highly dependent on the thermal loading and the phase state of the alloy. Based on these facts, the fatigue failure of these alloys can be classified into three different categories of (i) usual reduction in strength under cyclic loads, (ii) loss in functional properties, that is, superelasticity and shape memory effect, and (iii) change in material properties such as transformation temperatures under thermal cyclic loads. These three types of fatigue are called structural fatigue, functional fatigue, and thermal fatigue, respectively. Definitions and studies regarding each fatigue type are discussed in the following sections. It is worth mentioning that the majority of the studies concerning the fatigue of SMAs have been dedicated to NiTi.

5.2.1 Structural Fatigue

Similar to other materials, SMAs may experience a reduction in strength under repeated loads and consequently fail at a lower stress level than their nominal tensile strength. As stated before, the crack initiation approaches for fatigue behavior of SMAs are more popular; hence, the majority of the studies in this area follow the strain-life or stress-life methods for fatigue analysis. Nevertheless, the fatigue crack growth behavior of SMAs has been also taken into account in some experimental and analytical studies [28–30].

5.2.1.1 Uniaxial Fatigue Behavior

As mentioned before, uniaxial fatigue studies can be performed using either the force- or strain-controlled (displacement-controlled) method. Under strain-controlled tests, a cyclic hardening or softening may occur due to the dislocation density effects. Similarly, an increase or decrease in strain at a constant stress level may occur in force-controlled tests. For some shape memory materials such as NiTi alloys, using strain-controlled tests to evaluate the fatigue behavior is more appropriate than using the more traditional force-controlled tests. This can be concluded from the monotonic stress–strain curve of the material in different phases, presented in Figure 5.7, revealing a stress plateau for a wide range of strains, in both superelastic and shape memory behaviors.

A major challenge in the fatigue analysis of SMAs is the drastic effects of the chemical composition on the fatigue behavior. Such effects can be observed, as an example, in fatigue analysis of Cu–Al–Ni alloys [8]. As can be seen from Figure 5.8, three types of Cu–Al–Ni alloys, A, B, and C, with slightly different chemical compositions, listed in Table 5.1, exhibit significantly different fatigue behaviors as well as transformation temperatures. Similar to other metals, the strain-based and stress-based fatigue analysis of the same materials may lead to different results. As can be seen from Figure 5.8a, in stress-life

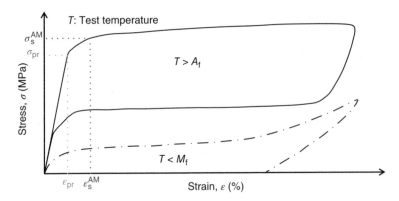

Figure 5.7 Schematic stress–strain curves of NiTi alloy in different phases (T represents the test temperature). Reproduced with permission from Ref. [31], Taylor & Francis

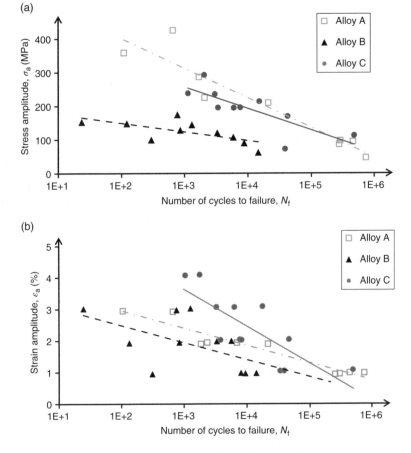

Figure 5.8 (a) Stress-life and (b) strain-life fatigue data for the same sets of Cu–Al–Ni data from strain-controlled tests [8]. Reproduced with permission from Ref. [10], John Wiley & Sons, Ltd

Table 5.1 Details of Cu–Al–Ni specimens in Sakamoto et al.'s study [8]

Alloy	Composition (mass %)	M_s (°C)
A	$Cu_{81.0\%}Al_{14.9\%}Ni_{4.1\%}$	−93
B	$Cu_{82.0\%}Al_{14.0\%}Ni_{4.0\%}$	0
C	$Cu_{83.0\%}Al_{13.2\%}Ni_{3.8\%}$	147

fatigue analysis, type A is the most resistant and type B is the least resistant to fatigue, while in strain-life fatigue analysis, that is, Figure 5.8b, type A specimens have the longest and type B specimens have the shortest fatigue lives. Since types A, B, and C were similar materials with slightly different chemical compositions, these results demonstrate the significant effects of the chemical composition on the fatigue behavior.

5.2.1.2 Stress-Life Fatigue Behavior

The stress-life approach is the most commonly used method for fatigue analysis of metallic materials. Nevertheless, due to some technical and practical reasons, the majority of fatigue studies on SMAs have followed the strain-controlled test procedures and strain-life approach. It should be noted that these alloys are often used for load-carrying applications; therefore, it seems necessary to also study the fatigue behavior employing the stress-life approach.

Two stress-life curves for NiTi are presented in Figure 5.9. This figure represents the fatigue stress-life data for specimens with the $A_f = 27°C$ at different test temperatures [32]. The $S–N$ curve is linear for higher temperature tests and bilinear for lower temperature tests where the test temperature is very close to the austenite finish temperature, A_f. These results clearly indicate that in the stress-life representation of fatigue data, the fatigue life increases with an increase in the test temperature above A_f. This increase can be mainly attributed to the higher stress plateau levels, obtained for the same material at higher temperatures. This higher stress plateau will increase the load-carrying capacity and improve the stress-based fatigue behavior of the material. Stress-life fatigue data presented in Figure 5.9 for NiTi SMAs may be divided into several linear sections. Furthermore, some stress-life experimental results, including the data presented in Figure 5.9, have revealed endurance limits for NiTi SMAs.

Stress-life fatigue curves for Cu–Zn–Al SMAs are presented in Figure 5.10. In this figure, the chemical compositions of alloys "B" and "C" are listed in

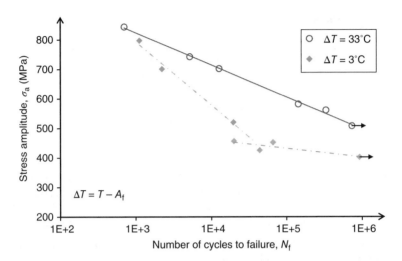

Figure 5.9 Stress-life fatigue data for a superelastic NiTi alloy. *T* is the test temperature and A_f is the material's austenite finish temperature. Arrows indicate that the failure did not occur in the specimen. From Ref. [32]

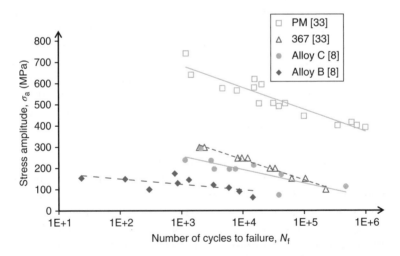

Figure 5.10 Stress-life data for different shape memory alloys from multiple studies. Reproduced with permission from Ref. [8], the Japanese Institute of Metals. Reproduced with permission from Ref. [33], Société française de physique, EDP Sciences

Table 5.1, "367" is a hot-rolled $Cu_{68.4}Zn_{27.5}Al_{4.2}$ alloy, and "PM" is an extruded $Cu_{76}Zn_{14.8}Al_{8.3}$ alloy. As can be seen from these data sets, the stress-life behavior for these SMAs can be presented linearly in a semilogarithmic plot. Considering the data presented in Figures 5.9 and 5.10, the stress-life fatigue behavior of SMAs may be expressed by a series of linear equations in semilogarithmic plots.

5.2.1.3 Strain-Life Fatigue Behavior

SMAs can undergo a large strain range at a constant stress level when the stress reaches a certain level, that is, the stress plateau (Figure 5.7). Consequently, the stress–strain relation is linear, only for a very small value of the strain, compared to the total recoverable strain. Due to this fact, the force-controlled tests are less accurate than the strain-controlled tests for these alloys. In addition, strain-based testing does not require very specialized equipment since they can be conducted using rotary-bending tests on wire specimens, as described in Figure 5.6. For these reasons, the majority of fatigue studies on SMAs, and specifically the NiTi alloys, are based on strain-life (ε–N) approach.

According to strain-life fatigue data for superelastic NiTi alloys (i.e., test temperatures above the A_f), presented in Figure 5.11, there is a reverse relation

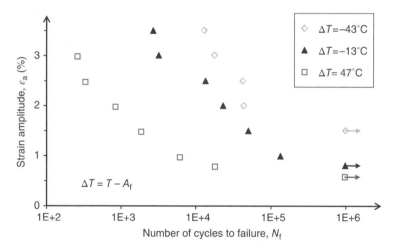

Figure 5.11 Effects of test temperature on the rotary-bending strain-life fatigue behavior of a NiTi shape memory alloy. T is the test temperature and A_f is the austenite finish temperature. Reproduced with permission from Ref. [34] of Elsevier

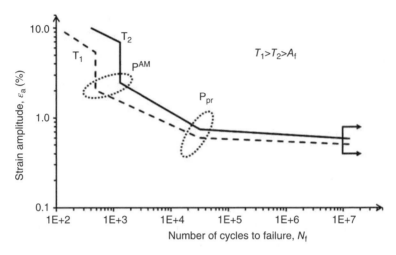

Figure 5.12 Schematic fatigue life regimes of a superelastic NiTi alloy in strain-controlled tests (T is the test temperature and A_f is the material austenite finish temperature) [25]

between the test temperature and the fatigue resistance of the material. Therefore, as the test temperature decreases, the fatigue life of the NiTi specimens increases in strain-life approach. Furthermore, it can be seen from this figure that as the number of cycles increase, the ε–N curves tend to converge around 10^5 cycles which implies insensitivity of the ε–N curve to temperature after a certain number of cycles.

Generally, the strain-life data show four distinct regimes for fatigue of superelastic NiTi at temperatures above A_f. This classification is comprehensively discussed by Pelton et al. [25] and schematically presented in Figure 5.12. The first regime in the ε–N plot is related to very high strain amplitudes ($\varepsilon_a \sim 5$–10%) which leads to a very low-cycle fatigue failure. In this region, fatigue life (in logarithmic form) decreases, almost linearly, with an increase in the strain amplitude. The fatigue life of NiTi alloys in this regime is highly sensitive to other test parameters such as surface finish [25] and test temperature [25, 34]. The second regime includes the data for high strain amplitudes ($\varepsilon_a \sim 2$–7%). In this region, the fatigue life seems to be independent of strain amplitude and is almost constant. The third regime includes the strain amplitudes between approximately 0.7 and 2%, where the fatigue life constantly decreases as the strain amplitude increases. The life of the specimens in this regime is found to be in the range of approximately 10^3

to 10^5 cycles. Pelton et al. [25] suggested that a modified Coffin–Manson relation [16, 17] with a coefficient that decreases with increasing test temperature can be used to represent the fatigue behavior in this regime. The fatigue behavior in the third regime is mainly related to the transition from the linear elastic behavior in austenite phase to the stress-induced martensitic phase (i.e., the stress plateau of the superelastic stress–strain plot, shown in Figure 5.7). The fourth regime includes the high-cycle fatigue data and is very similar to classical fatigue for linear elastic metals. In this regime, Basquin's model [15] can be used to represent the fatigue behavior. The second turning point in fatigue curves (i.e., Point P^{AM} in Figure 5.12) of a superelastic NiTi alloy is related to the starting point of the austenite-to-martensite transformation (i.e., ε_S^{AM} in Figure 5.7), and the third turning point (i.e., point P_{pr} in Figure 5.12) corresponds to the proportional limit (i.e., ε_{pr} in Figure 5.7) of this material [34].

For the shape memory NiTi alloy, no such distinct regimes may be observed in fatigue data. The data for this type of alloys can be classified in two regions, for each the behavior is approximately linear in a log–log scale. It is notable that no distinct low-cycle fatigue regimes similar to the ones typically seen in superelastic alloys can be observed in the shape memory NiTi specimens.

5.2.1.4 Mean Stress/Strain Effects on Fatigue Behavior

In the previous section, the fatigue behavior of SMAs under fully reversed loading was discussed. Nevertheless, in many applications, SMAs may be subjected to a mean strain or stress during the life of the component. Actuators made of NiTi wires, as discussed in Chapter 3, are good examples of shape memory applications under nonzero mean strain cyclic loads. In these applications, the wire is under a minimum strain, which is usually greater than zero, and a maximum strain, during the actuation cycle. As another example of nonzero mean stress, NiTi superelastic stents are subjected to two types of loads: a tension from the artery and a cyclic load due to the pulsation of the artery wall caused by the changes in systolic–diastolic of the blood pressure.

Mean stresses significantly affect the fatigue behavior of SMAs. Fatigue testing of these alloys under tensile mean stresses indicates a drastic reduction in the fatigue strength. However, this behavior exists mainly for the stresses below the proportional limit. Fatigue resistance of superelastic alloys reduces considerably in the presence of mean tensile stress as long as the maximum stress is not high enough to induce the martensitic phase transformation [35]. The S–N curves from identically treated dogbone-shaped NiTi alloys under tension–compression loading demonstrate the reducing effects of

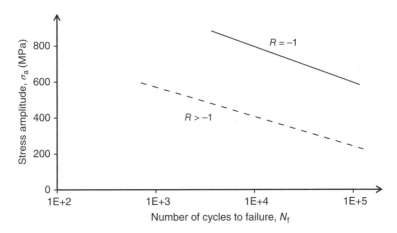

Figure 5.13 A schematic representation of mean stress effects on the fatigue behavior of superelastic NiTi dogbone specimens. R indicates the load ratio, $R = \sigma_{min}/\sigma_{max}$

tensile mean stresses on the fatigue resistance as presented in Figure 5.13 [35]. Again, it should be emphasized that the reducing effects of the tensile mean stress on the fatigue resistance only exist while the mean stress is not large enough to cause stress-induced martensitic phase transformation. Experimental results with different mean stresses indicate the possible beneficial effects of tensile mean stresses on the fatigue of superelastic NiTi by inducing the martensitic phase transformation [36]. Accordingly, the classical formulation for the tensile mean stress effects on the fatigue of metallic materials, such as Goodman and Soderberg equations [1, 19], are not applicable in such cases.

Similar to the mean stress effects, there is not a consistent effect of mean strain on the fatigue behavior of SMAs and the effects may depend on the mean strain level. One common method to compare the results for different combination of mean strains and strain amplitudes is to compare these combinations at a certain run-out life (e.g., 10^6 or 10^7 cycles). In this method, experimental results from different combinations of mean and alternating strains are plotted together and compared. Examples of this method, presenting the mean strain effects on the fatigue behavior of NiTi, from the experiments conducted by Tolomeo et al. [21] for 10^6 cycles run-out life and Pelton et al. [37] for 10^7 cycles run-out life alloys are shown in Figure 5.14. In this plot, the vertical axis is the alternating strain (i.e., strain amplitude), ε_a, and the horizontal axis is the mean strain, ε_m. According to the data reported by Tolomeo et al. [21], the tolerable alternating strain increases with increasing mean strain up

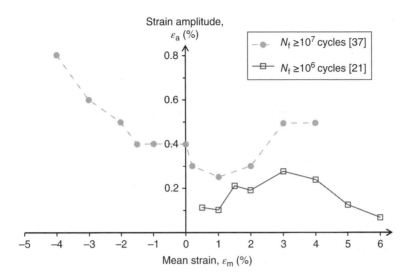

Figure 5.14   Mean strain effects on fatigue behavior of superelastic NiTi alloys

to a certain mean strain amplitude (e.g., 3%). Beyond 3% mean strain, however, the tolerable alternating strain, without failure of the specimen, decreases as the mean strain increases.

Analogous to the data from Tolomeo et al. [21], the observations by Pelton et al. [37] indicate a similar behavior for effects of mean strain on the fatigue behavior of superelastic NiTi alloys for mean strains lower than 4.0%, as presented in Figure 5.14. The latter study also includes compressive mean strains which are calculated using finite element analysis on the stent-like components. Their findings for superelastic NiTi alloys indicate that the bearable alternating strain decreases as the mean strain increases up to 1%. Above 1% mean strain, the alternating strain tends to increase by an increased mean strain level. This behavior can be attributed to the formation of stress-induced martensitic phase in superelastic NiTi alloys under higher mean strain values [37]. Similar results were also reported by Morgan and coworkers [38] indicating beneficial effects of stress-induced martensitic phase transformation resulted from larger mean strains.

The classical mean stress/strain correction methods such as Goodman or Soderberg relations are not appropriate for superelastic NiTi alloys [36, 38, 39]. In addition, fatigue resistance of NiTi alloys may not always decrease by an increase in mean stress or strain.

5.2.1.5 Torsional Fatigue

Torsional load, similar to tension, compression, and bending, is one of the typical loading conditions for components and structures made of SMAs [40–42]. Unlike rotary-bending and uniaxial fatigue, the fatigue of these alloys under torsional loads is generally very similar to the typical fatigue curves for other metallic materials. In other words, the very distinct multilinear behavior of SMAs is not typically observed for the torsional fatigue in a semilogarithmic presentation of the test results. The general reduction in the strength of the material is clearly observed under torsion in both stress-life and strain-life fatigue analysis as can be seen in Figure 5.15a and b, respectively. Specimen type may affect the torsional fatigue of NiTi alloys. Similar to the observation reported for the effect of cross-sectional shape on the torsional fatigue of mild steel [43, 44], higher torsional fatigue lives have been reported in the literature for specimens with solid cross section compared to those with hollow cross sections, as illustrated in Figure 5.16.

5.2.1.6 Phase Contribution to Fatigue Behavior

Effects of the microstructural phase to the fatigue behavior of the NiTi SMAs can be realized by testing identical samples under various test temperatures, as the test temperature dictates the phase of the material. Fatigue testing technique (i.e., force-controlled or strain-controlled) and the method employed to present the fatigue data (i.e., stress-life or strain-life) can illustrate different effects of the material phase on the fatigue. For stress-based fatigue analysis and stress-life fatigue data, the superelastic NiTi alloys show higher fatigue strength compared to the shape memory NiTi alloys, as illustrated in Figure 5.17 [5]. In this figure, the superelastic specimens, for which the test temperature was higher than the austenite finish temperature, exhibited higher fatigue strength compared to the specimens in mixed austenitic–martensitic phase. However, in strain-based presentation of the fatigue data, the behavior is completely different, and shape memory NiTi alloys show significantly higher fatigue strength, as presented in Figure 5.11. For the NiTi alloy of Figure 5.11, the transformation temperatures A_f and M_f are approximately 77°C and −11°C, respectively. Accordingly, for the tests with negative ΔT, $\Delta T = T - A_f$, specimens are in mixed phase, while for $\Delta T > 0$, specimens are superelastic. The seemingly contradictory results in different phases can be attributed to the fatigue life approach employed to analyze the data. In summary, austenitic (superelastic) NiTi exhibits higher fatigue resistance in stress-based fatigue approach, whereas martensitic (shape memory) NiTi alloys typically have longer lives in strain-based approach. The higher

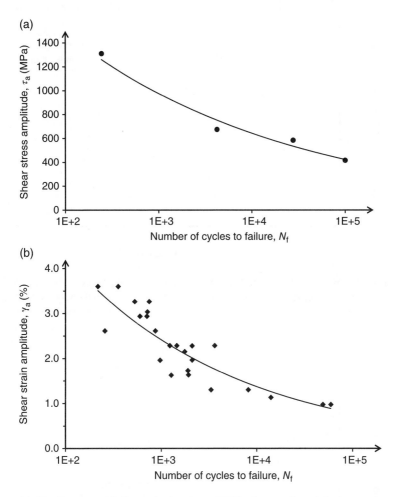

Figure 5.15 Torsional fatigue behavior of NiTi alloys using (a) shear stress-life approach [40] and (b) shear strain-life approach. Reproduced with permission from Ref. [42], Elsevier

stress-life fatigue resistance of the superelastic NiTi alloy can be attributed to the higher load-carrying capacity of the superelastic NiTi alloys as compared to the martensitic NiTi alloys, as illustrated in Figure 5.7.

The effect of material phase on the fatigue resistance of superelastic alloys is also significant. Phase transformation may take place in superelastic NiTi components in the presence of mean strains, which can accelerate or decelerate the fatigue crack initiation process. An increase in the mean strain in

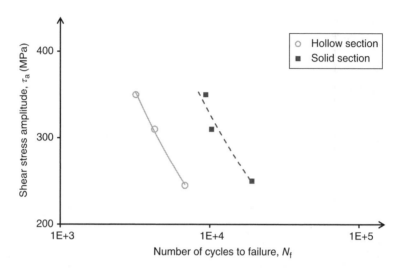

Figure 5.16 Effects of the specimen's cross section type (i.e., solid vs. hollow) on the fatigue behavior of NiTi alloys. Reproduced with permission from Ref. [41], Elsevier

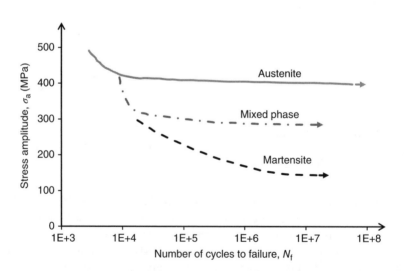

Figure 5.17 Effects of material phase in stress-life fatigue analysis of NiTi alloys. Reproduced with permission from Ref. [5], Elsevier

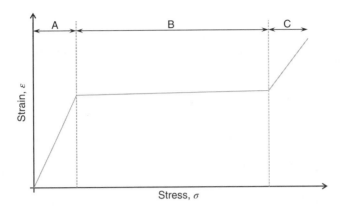

Figure 5.18   Multiple phases for superelastic NiTi alloy; A, linear elastic region; B, stress plateau region; and C, detwinned martensitic region. Beyond region C, the material NiTi alloy behaves similar to other metals and may undergo a plastic strain

the austenite phase, which is up to the linear elastic limit (i.e., region A in Figure 5.18), is detrimental to the fatigue strength. However, an increase in the mean strain beyond the linear elastic limit in the plateau (i.e., region B in Figure 5.18) is beneficial to the fatigue resistance as long as the total applied strain remains within the superelastic region (i.e., regions B and C in Figure 5.18). The formation of stress-induced martensite in the superelastic plateau region is the main reason for the improved fatigue strength in superelastic NiTi alloys [6]. Finally, an increase in mean strain level in the detwinned stress-induced martensitic phase (i.e., region C in Figure 5.18) results in a reduced fatigue resistance. It is worth mentioning that beyond region C, the NiTi alloy behaves similar to other metals and plastic strains may be observed.

For superelastic NiTi alloys, as long as the crack nucleation is inhibited, the stress-induced martensite (i.e., region B in Figure 5.18) is the preferred condition. In this region, the superelastic alloy demonstrates a higher crack propagation resistance when compared to the fully austenitic phase (i.e., region A in Figure 5.18). However, shape memory NiTi alloys usually exhibit drastically superior resistance to fatigue crack propagation than the superelastic alloys [28, 45].

5.2.2 Functional Fatigue

The term "functional fatigue" was first proposed by Eggeler and coworkers in 2004 [24] indicating a decrease in the functionality of the SMAs during cyclic

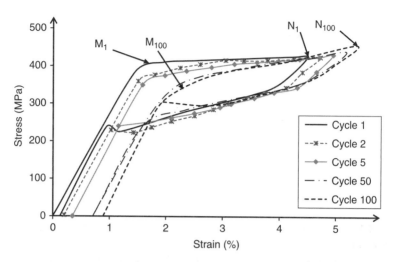

Figure 5.19   Schematic representation of the cyclic change in the stress-strain curve of a superelastic NiTi alloy, indicating a reduction in the starting stress-induced martensitic phase, as well as the area of the hysteresis loop

thermomechanical loading. For superelastic NiTi alloys, there is a reduction in martensite-inducing stress level (i.e., points indicated by M in Figure 5.19) with an increase in the number of cycles, as illustrated in Figure 5.19. In addition, as the number of cycles increases, the maximum recoverable strain decreases. Finally, the area of the hysteresis loop of a superelastic NiTi alloys decreases with an increase in number of cycles, yielding to a narrow stable hysteresis loop [35]. All these changes can be classified as functional fatigue since they deal with the functional property of SMAs. The area of the hysteresis loop indicates the capability of the material to absorb and dissipate energy. The decrease in the maximum recoverable strain is also a functional defect since the superelastic NiTi alloys in most applications undergo a large recoverable deformation. The decrease in the maximum recoverable strain is due to the formation of residual local plastic strains which can cause slip in the crystalline structure.

The changes in the functional properties of a superelastic NiTi alloy follow two main trends. First, the rate of the change is faster in the very first few cycles and decreases as the number of cycles increases. Second, the functional property reaches a stable state after a limited number of cycles, that is, approximately 100–150 cycles. This is the main reason that the alloys are taken through cyclic thermomechanical loadings to induce stable functional properties before using in a device.

The functional fatigue under multiaxial cyclic loading is also possible. Similar to uniaxial loading, the energy dissipation capacity of superelastic NiTi alloys decreases for the first few cycles and reaches a stable state [46]. The behavior of the Cu–Zn–Al SMAs depends on whether the material is a single crystalline or polycrystalline. In a single-crystalline alloy, in which the crystal lattice of the entire sample is continuous, there is a significant increase in the area of the hysteresis loops with increasing loading cycles, which indicates an increase in energy dissipation capacity. The corresponding stress to the ultimate strain also increases significantly for single-crystalline Cu–Zn–Al alloys [47]. In contrast to single-crystalline specimens, experimental observations for the polycrystalline Cu–Zn–Al alloys indicate a reduction in the area of hysteresis loop by an increase in load cycles [48].

5.2.3 Thermal Fatigue

The behavior of SMAs under cyclic thermal loading has been of great interest since SMAs were first synthesized. Several studies have been performed to evaluate the behavior of these materials under several cycles of thermal loading with or without application of external mechanical loads. Although it is obvious that the thermal cycling has great influence on the behavior of SMAs, the extension of these effects are not yet completely understood. This problem is due to the diversity of SMAs and variations in material properties with slight changes in the chemical composition. Contradictory effects are reported on the effects of thermal cycles on the martensite start temperature, M_s. Some studies have reported a decrease in M_s [49], while others indicate an increase in M_s under thermal cycles [50, 51]. For other phase characteristic temperatures such as A_s, A_f, and M_f, there have also been some inconsistencies observed in the experimental data. Drastic changes in the transformation temperatures of the NiTi SMA may occur under thermally cyclic loading and, generally, a reduction in the transformation temperatures of NiTi alloys has been reported in the literature. Thermomechanical cycling (i.e., thermal cycling in presence of external mechanical loads) may also greatly influence the functional properties of the SMAs, reducing the hysteresis loop as well as the recoverable strain magnitude.

5.3 Microstructural Effects

Microstructure is defined as the structure of a prepared surface or a thin foil of the material as revealed by magnification under microscope. The microstructure of a material may strongly influence mechanical properties such as tensile and fatigue strengths, which consequently may also influence the functionality of the component.

The microstructural features of a material including grain size, lattice defects, defect volume fraction, inclusions, boundary misorientation, and the grain crystallographic orientation are influenced by chemical compositions as well as manufacturing and postmanufacturing processes. Finer grain sizes generally improve the high-cycle fatigue resistance of the material due to fewer stress concentrations on the grain boundaries [33]. For some SMAs such as Ti–Mo-based alloys, the stress value for inducing the martensitic phase may increase by a reduction in grain size and an increase in the dislocation density [52].

The difference in grain size is one of the major sources for different fatigue lives observed for Cu–Zn alloys and NiTi alloys under cyclic loads. The grain size mainly influences the stress concentration on the grain boundaries which may cause damage in the stress-induced martensitic SMAs. Stress concentrations at the grain boundaries eventually lead to permanent deformation and crack initiation. For materials with finer grain size, the localized strains along the slip bands seem to be reduced. Moreover, microcracks tend to initiate from the locations with more stress concentrations, such as inclusions, laminations, and voids/porosities under cyclic loadings or thermal cycling.

5.4 Treatment and Postprocessing Effects

One of the key parameters that can considerably influence the fatigue resistance of SMAs is the final treatment of the material. Based on transmission electron microscopy (TEM) observations, Pelton concluded that "processing plays a key role in the fatigue behavior of Nitinol" [6].

Several different methods are proposed in the literature to construct a more stable SMA using different treatment techniques. The combination of cold work and aging treatments may lead to an optimized state of microstructure which consequently results in a more stable SMA under cyclic loading. As this combination results in a minimized amount of unresolved strains, it is preferred over the combination of annealing and age treatment for applications involving cyclic loading.

5.5 Other Influential Parameters

In addition to the influential parameters discussed previously in this chapter, there are some other parameters that may also affect the fatigue behavior of SMAs such as test temperature, surface finish, specimen geometry, load

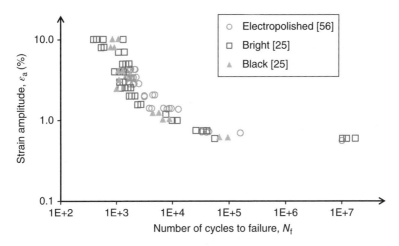

Figure 5.20 Effects of surface finish quality on the fatigue behavior of NiTi shape memory alloys

frequency (rotational speed in rotary-bending test), and material impurities. Effects of these parameters on the fatigue behavior of SMAs have occasionally been studied in the literature [53–55]. In this section, some of these influential parameters are briefly discussed.

The test temperature greatly affects the fatigue resistance of SMA components. Representations of the effects of the test temperature on fatigue behavior of NiTi are presented in Figures 5.11 and 5.12. NiTi alloys demonstrate higher fatigue lives for lower test temperatures under strain-life fatigue analysis. As presented in Figure 5.11, in strain-based fatigue analysis, there is a reverse relation between the test temperature and the fatigue lives for superelastic NiTi alloys [25, 34].

The results from different types of surface finish for a similar composition and treatment of NiTi SMAs are presented in Figure 5.20 [25]. As can be seen in this figure, the effect of the surface finish is more significant in low-cycle fatigue and is not very considerable in high-cycle fatigue region. The electropolished specimens show slightly higher fatigue strength in the low-cycle fatigue regime as compared to specimens with bright or black surface finish.

Introducing additive materials to SMAs may not significantly affect the fatigue behavior of these alloys (e.g., adding Cu or TiC to typical NiTi alloys). As an example, Vaidyanathan et al. [30] studied the effects of TiC,

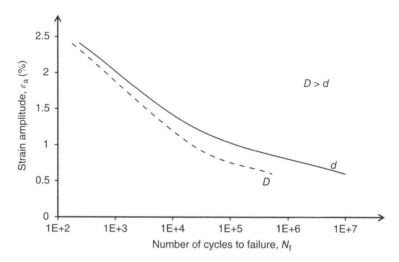

Figure 5.21 Effects of specimen's diameter on the fatigue behavior of NiTi shape memory alloy wires, d and D indicate wire diameter and $D > d$

as reinforcement, on fatigue crack growth properties of a NiTi SMA by using two different contents of TiC: 10 and 20%. They concluded that the overall fatigue behaviors of the reinforced SMAs are similar to the original alloy and there is no major change corresponding to the amount of TiC, even though the uniaxial stress–strain behaviors of the specimens are quite different. However, a slight reduction in the fatigue strength of NiTi alloys may be expected by adding 10% Cu to the chemical composition of these SMAs [34].

Since most of the studies on the fatigue behavior of SMAs are performed on wire specimens using rotary-bending test setups, the effects of the wire diameter on the fatigue behavior need to be considered. Figure 5.21 schematically presents the strain-life fatigue behaviors of NiTi wires with different diameters. As can be seen from this figure, some effects of specimen size on the fatigue behavior of NiTi alloys may be expected. Higher fatigue strengths are reported in the literature for smaller diameter wires [53, 57]. However, Yao and coworkers [57] have reported the cross-sectional shape to be a more important factor on the fatigue behavior of NiTi alloys as compared to the specimen diameter. Although more data is required to generalize the effects of specimen size on the fatigue behavior of SMAs, lower fatigue lives for larger specimens may be expected as more impurities typically exist on larger cross-sectional areas.

5.6 Examples

In this section, few examples are provided to explain the practical aspects of the fatigue analysis and its applications to design. Example problems are designed so that they cover many real applications of the SMAs.

Example 5.1. The strain-life fatigue data of three different NiTi alloys are shown in Figure 5.22 at room temperature. Determine the appropriate material for the low-cycle fatigue application ($N_f < 10^3$), where the material is subjected to a large amount of strain. What material would you select for a high-cycle fatigue application ($N_f > 10^5$)?

Answer. As can be seen from Figure 5.22, the Type 3 NiTi alloy exhibits higher fatigue resistance in the low-cycle fatigue regime, for example, $\varepsilon_a = 9\%$. However, the Type 2 NiTi alloy in this figure shows superior fatigue resistance in the high-cycle fatigue regime, for example, $\varepsilon_a = 0.9\%$. Therefore, for applications in which the material is exposed to low-cycle fatigue (i.e., higher strain levels), Type 3 would be a better candidate, while Type 2 is more appropriate for high-cycle fatigue applications (i.e., lower strain levels).

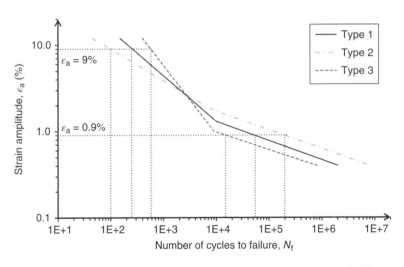

Figure 5.22 Strain-life fatigue data for three different types of NiTi alloy

Example 5.2. Assume a superelastic NiTi wire is subjected to an alternating fully reversed strain-controlled condition. If the fatigue behavior of the material follows the trend shown in Figure 5.23:

(a) What would be the expected fatigue life of the material at temperature T_2 for an alternating strain of 0.8%?
(b) Find the alternating strain that the wire can tolerate without failure for a design life of 10^4 cycles at the test temperature of $T = T_1$.

Answer. As is implied by this example, due to the lack of data and general predictive equations for the fatigue of NiTi alloys, there should be some case-specific fatigue data to be used in accordance to each alloy. For the same reason, we assume that the fatigue of the material follows the trend of an existing set of data.

(a) According to the fatigue curves, shown in Figure 5.23, at $T = T_2$, the life of the specimen corresponding to strain amplitude of 0.8% is about 30 000 cycles.
(b) At $T = T_1$, the corresponding strain value for the $N = 10^4$ from the curve is equal to $\varepsilon_a = 0.85\%$ (Figure 5.23).

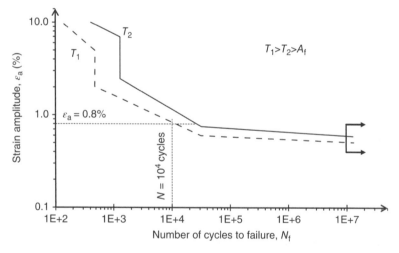

Figure 5.23 Strain-life fatigue data for a superelastic NiTi alloy at two different temperatures

Example 5.3. Assume the strain-life data for the NiTi wire used in the actuator shown in Figure 5.24a follows the plot shown in Figure 5.24b. Calculate the expected fatigue life of the wire if the maximum applied strain in each cycle of loading is 4%.

(a)

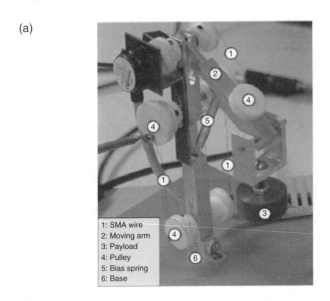

(b)

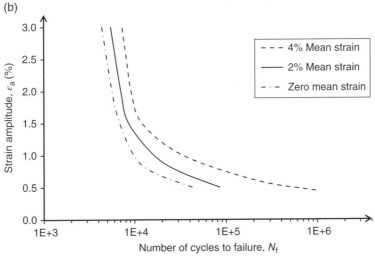

Figure 5.24 (a) A one degree-of-freedom rotary shape memory alloy actuator and (b) the strain-life fatigue data for the superelastic NiTi alloy used in this actuator

Answer. Considering the fact that actuators are typically under nonzero mean strain, we can assume the status of loading in the actuator follows a pulsating behavior ($R = \varepsilon_{min}/\varepsilon_{max} = 0$). If the maximum strain applied to the wire in each cycle is 4%, we can calculate the following parameters (see Figure 5.1):

$$\varepsilon_{max} = 4\%, \; \varepsilon_{min} = 0\% \; \rightarrow \; \varepsilon_m = \frac{(\varepsilon_{max} + \varepsilon_{min})}{2} = \frac{4\% + 0\%}{2} = 2\%$$

$$\varepsilon_a = \frac{(\varepsilon_{max} - \varepsilon_{min})}{2} = \frac{4\% - 0\%}{2} = 2\%$$

Using the data for 2% mean strain (i.e., solid curve in Figure 5.24b) with an alternating strain of 2%, the expected fatigue life of the wire will be approximately 8000 cycles.

Example 5.4. The fatigue data of a NiTi wire is presented by a circle and dashed line in Figure 5.14. In an actuator-type application, the wire is subjected to 0.4% alternating strain as well as a minimum strain of 1.6% resulted from the bias spring and gravity.

(a) Determine whether the wire endures 10^7 cycles of such loading.
(b) Propose an economical solution to improve the fatigue life of this actuator.

Answer. In order to use the data presented in Figure 5.14, the following parameters need to be calculated:
$\varepsilon_{min} = 1.6\%, \; \varepsilon_a = 0.4\% \; \rightarrow \; \varepsilon_m = \varepsilon_{min} + \varepsilon_a = 1.6\% + 0.4\% = 2\%$ (see Figure 5.1)

(a) According to data presented in Figure 5.14, the maximum tolerable alternating strain for the superelastic NiTi alloy with $\varepsilon_m = 2\%$ and a minimum fatigue life of 10^7 cycles is 0.3%. This value is smaller than the required alternating strain (0.4%) for this actuator; thus, the system will fail before 10^7 cycles.

(b) Several solutions can be proposed to improve the fatigue life of the wire, such as employing a stronger wire with higher microstructural quality and consequently higher fatigue life. However, considering the data presented in Figure 5.14, the most economical solution seems to be increasing the mean strain. Therefore, by selecting the bias spring so that the minimum strain on the wire increases to 2.6%, while keeping the alternating strain the same (which actually dominates the application of the actuator), the strain components on the wire will change to

$$\varepsilon_{min} = 2.6\%, \; \varepsilon_a = 0.4\% \; \rightarrow \; \varepsilon_m = \varepsilon_{min} + \varepsilon_a = 2.6\% + 0.4\% = 3\%$$

As can be seen from Figure 5.14, the maximum alternating strain for a minimum fatigue life of 10^7 cycles is 0.5%. This value is higher than the required alternating strain for this actuator, and therefore, the NiTi wire can withstand the desired life of 10^7 cycles.

References

[1] Stephens RI, Fatemi A, Stephens RR, Fuchs HO. Metal fatigue in engineering. New York: John Wiley & Sons, Inc.; 2000.

[2] Bigeon MJ, Morin M. Thermomechanical study of the stress assisted two way memory effect fatigue in TiNi and CuZnAl wires. Scr Mater 1996;35:1373–8.

[3] Rachinger WA. A "super-elastic" single crystal calibration bar. Br J Appl Phys 1958;9:250.

[4] Gambarini G. Cyclic fatigue of nickel–titanium rotary instruments after clinical use with low- and high-torque endodontic motors. J Endod 2001;27:772–4.

[5] Melton KN, Mercier O. Fatigue of NiTi thermoelastic martensites. Acta Metall 1979;27:137–44.

[6] Pelton AR. Nitinol fatigue: a review of microstructures and mechanisms. J Mater Eng Perform 2011;20:613–7.

[7] Plotino G, Grande NM, Cordaro M, Testarelli L, Gambarini G. A review of cyclic fatigue testing of nickel–titanium rotary instruments. J Endod 2009;35:1469–76.

[8] Sakamoto H, Kijima Y, Shimizu K. Fatigue and fracture characteristics of polycrystalline Cu–Al–Ni shape memory alloys. Trans Jpn Inst Metal 1982;23:579.

[9] Strnadel B, Ohashi S, Ohtsuka H, Miyazaki S, Ishihara T. Effect of mechanical cycling on the pseudoelasticity characteristics of Ti–Ni and Ti–Ni–Cu alloys. Mater Sci Eng A 1995;203:187–96.

[10] Yared GM, Dagher FE, Machtou P. Cyclic fatigue of ProFile rotary instruments after clinical use. Int Endod J 2000;33:204–7.

[11] Wilkes KE, Liaw PK. The fatigue behavior of shape-memory alloys. JOM 2000;52:45–51.

[12] Mahtabi MJ, Shamsaei N, Mitchell MR. Fatigue of Nitinol: The state-of-the-art and ongoing challenges. J Mech Behav Biomed Mater 2015;50:228–254.

[13] Robertson SW, Pelton AR, Ritchie RO. Mechanical fatigue and fracture of Nitinol. Int Mater Rev 2012;57:1–37.

[14] ASTM Standards. Standard terminology relating to fatigue and fracture testing. West Conshohocken, PA: ASTM International; 2000.

[15] Basquin OH. The exponential law of endurance tests. Proc Am Soc Test Mater, 1910;10:625–30.

[16] Coffin LF, Tavernelli JF. The cyclic straining and fatigue of metals. Trans Met Soc AIME 1959;215:794–807.

[17] Manson SS, Dolan TJ. Thermal stress and low cycle fatigue. J Appl Mech 1966;33:957.

[18] Soderberg CR. Factor of safety and working stress. Trans ASME 1939;**52**:13–28.

[19] Budynas RG, Nisbett JK, Shigley JE. Shigley's mechanical engineering design. New York: McGraw-Hill; 2011.

[20] Pelton AR, Schroeder V, Mitchell MR, Gong X-Y, Barney M, Robertson SW. Fatigue and durability of Nitinol stents. J Mech Behav Biomed Mater 2008; **1**:153–64.

[21] Tolomeo D, Davidson S, Santinoranont M. Cyclic properties of superelastic Nitinol: design implications. SMST-2000 Proceedings of the International Conference on Shape Memory Superelastic Technologies, Pacific Grove, CA, USA, April 30– May 4, 2000. Russell SM, Pelton AR, eds, Fremont, CA: SMST, the International Organization on Shape Memory and Superelastic Technology; 2000, p. 471.

[22] ASTM Standards. Practice for conducting force controlled constant amplitude axial fatigue tests of metallic materials. West Conshohocken, PA: ASTM International; 2007.

[23] ASTM Standards. Test method for strain-controlled fatigue testing. West Conshohocken, PA: ASTM International; 2012.

[24] Eggeler G, Hornbogen E, Yawny A, Heckmann A, Wagner M. Structural and functional fatigue of NiTi shape memory alloys. Mater Sci Eng A 2004;**378**:24–33.

[25] Pelton AR, Fino-Decker J, Vien L, Bonsignore C, Saffari P, Launey M, Mitchell MR. Rotary-bending fatigue characteristics of medical-grade Nitinol wire. J Mech Behav Biomed Mater 2013;**27**:19–32.

[26] Tobushi H, Hachisuka T, Yamada S, Lin P-H. Rotating-bending fatigue of a TiNi shape-memory alloy wire. Mech Mater 1997;**26**:35–42.

[27] ASTM E2948-14. Standard test method for conducting rotating bending fatigue tests of solid round fine wire. West Conshohocken, PA: ASTM International; 2014.

[28] McKelvey AL, Ritchie RO. Fatigue-crack growth behavior in the superelastic and shape-memory alloy Nitinol. Metall Mater Trans A 2001;**32**:731–43.

[29] Robertson SW, Ritchie RO. In vitro fatigue-crack growth and fracture toughness behavior of thin-walled superelastic Nitinol tube for endovascular stents: a basis for defining the effect of crack-like defects. Biomaterials 2007;**28**:700–9.

[30] Vaidyanathan R, Dunand DC, Ramamurty U. Fatigue crack-growth in shape-memory NiTi and NiTi–TiC composites. Mater Sci Eng A 2000;**289**:208–16.

[31] Pelton AR, Dicello J, Miyazaki S. Optimisation of processing and properties of medical grade Nitinol wire. Minim Invasive Ther Allied Technol 2000;**9**:107–18.

[32] Miyazaki S, Sugaya Y, Otsuka K. Effects of various factors on fatigue life of Ti–Ni alloys. Proc MRS Int Meet Adv Mater 1988;**9**:251–6.

[33] Janssen J, Willems F, Verelst B, Maertens J, Delaey L. The fatigue properties of some Cu–Zn–Al shape memory alloys. J Phys Colloq 1982;**43**:C4–809.

[34] Miyazaki S, Mizukoshi K, Ueki T, Sakuma T, Liu Y. Fatigue life of Ti-50 at.% Ni and Ti-40Ni-10Cu (at.%) shape memory alloy wires. Mater Sci Eng A 1999; **2739r** S:658–63.

[35] Moumni Z, Herpen AV, Riberty P. Fatigue analysis of shape memory alloys: energy approach. Smart Mater Struct 2005;**14**:S287.

[36] Tabanli RM, Simha NK, Berg BT. Mean stress effects on fatigue of NiTi. Mater Sci Eng A 1999;**2739r S**:644–8.

[37] Pelton AR, Gong X-Y, Duerig T. Fatigue testing of diamond-shaped specimens. Med Device Mater-Proc Mater Process Med Devices Conf, 2003;**2003**:199–204.

[38] Morgan N, Painter J, Moffat A. Mean strain effects and microstructural observations during in vitro fatigue testing of NiTi. SMST-2003 Proceedings of the International Conference on Shape Memory and Superelastic Technologies, May 16–20, Pacific Grove, CA, USA. Pelton AR, Duerig T, eds, Menlo Park, CA: SMST Society Inc.; 2004, p. 303–10.

[39] Tabanli RM, Simha NK, Berg BT. Mean strain effects on the fatigue properties of superelastic NiTi. Metall Mater Trans A 2001;**32**:1866–9.

[40] Jensen DM. Biaxial fatigue behavior of NiTi shape memory alloy. Air Force Institute of Technology, Wright–Patterson Air Force Base, Fairborn, OH; 2005.

[41] Predki W, Kledki M, Knopik A. Cyclic torsional loading of pseudoelastic NiTi shape memory alloys: damping and fatigue failure. Mater Sci Eng A 2006; **417**:182–9.

[42] Runciman A, Xu D, Pelton AR, Ritchie RO. An equivalent strain/Coffin pseudoelastic ies of superela fatigue and life prediction in superelastic Nitinol medical devices. Biomaterials 2011;**32**:4987–93.

[43] Shamsaei N, Fatemi A. Effect of hardness on multiaxial fatigue behaviour and some simple approximations for steels. Fatigue Fract Eng Mater Struct 2009; **32**:631–46.

[44] Shamsaei N, Fatemi A. Deformation and fatigue behaviors of case-hardened steels in torsion: experiments and predictions. Int J Fatigue 2009;**31**:1386–96.

[45] Holtz RL, Sadananda K, Imam MA. Fatigue thresholds of Ni–Ti alloy near the shape memory transition temperature. Int J Fatigue 1999;**21**, Supplement 1: S137–S145.

[46] Song D, Kang G, Kan Q, Yu C, Zhang C. Non-proportional multiaxial transformation ratchetting of super-elastic NiTi shape memory alloy: experimental observations. Mech Mater 2014;**70**:94–105.

[47] Patoor E, Siredey N, Eberhardt A, Berveiller M. Micromechanical approach of the fatigue behavior in a superelastic single crystal. J Phys IV 1995;**5**:C8-227–232.

[48] Sade M, Hornbogen E. Thermal and mechanical fatigue of shape memory alloys. European Symposium on Martensitic Transformations. EDP Sciences; 1989, p. 125–32.

[49] Tadaki T, Takamori M, Shimizu K. Thermal cycling effect in Cu–Zn–Al shape memory alloys with B2 and DO$_3$ type ordered structures in parent phase. Trans Jpn Inst Met 1987;**28**:120–8.

[50] Perkins J, Muesing WE. Martensitic transformation cycling effects in Cu–Zn–Al shape memory alloys. Metall Trans A 1983;**14**:33–6.

[51] Sade M, Ahlers M. Low temperature fatigue in Cu–Zn–Al single crystals. Scr Metall 1985;**19**:425–30.

[52] Grosdidier T, Combres Y, Gautier E, Philippe M-J. Effect of microstructure variations on the formation of deformation-induced martensite and associated

tensile properties in a β metastable Ti alloy. Metall Mater Trans A 2000; **31**:1095–106.

[53] Norwich DW, Fasching A. A study of the effect of diameter on the fatigue properties of NiTi wire. J Mater Eng Perform 2009;**18**:558–62.

[54] Reinoehl M, Bradley D, Bouthot R, Proft J. The influence of melt practice on final fatigue properties of superelastic NiTi wires. SMST-2000: Proceedings of the International Conference on Shape Memory and Superelastic Technologies, Pacific Grove, CA, USA, April 2000, p. 397–403.

[55] Sawaguchi TA, KaustrchiG G, Yawny A, Wagner M, Eggeler G. Crack initiation and propagation in 50.9 at.% Ni–Ti pseudoelastic shape-memory wires in bending-rotation fatigue. Metall Mater Trans A 2003;**34**:2847–60.

[56] Sheriff J, Pelton AR, Pruitt LA. Hydrogen effects on Nitinol fatigue. SMST-2004 In Mertmann M, ed. Proceedings of the International Conference on Shape Memory and Superelastic Technologies, October 3–7, 2004, Baden-Baden, Germany. ASM International, p. 111–6.

[57] Yao JH, Schwartz SA, Beeson TJ. Cyclic fatigue of three types of rotary nickel–titanium files in a dynamic model. J Endod 2006;**32**:55–7.

6

Fabricating NiTi SMA Components

Christoph Haberland and Mohammad H. Elahinia

While there are several common processing steps for manufacturing shape memory and superelastic NiTi products, there is no single recipe to make these devices. Usually, there is a series or combination of different and individually selected processes, and the processing route depends on the desired shape, the desired properties, and the desired application. It is well known that the structural and functional properties of NiTi are highly sensitive to process parameters and steps. As a result, each process or process step in the production of semifinished or final NiTi components requires a considerable effort to account for these sensitive properties. Hence, a deep understanding of each process and its effects on product properties and performance is crucial for processing high-quality NiTi products.

This chapter reviews the most common, relevant, and established NiTi processing methods including conventional metallurgy, powder metallurgy (PM), forming, machining and cutting, joining, heat treatments, and finishing processes. Also, novel processing approaches are included in this chapter. These promising new processes are expected to receive more attention toward commercialization of niche devices as well as for industrial manufacturing.

Shape Memory Alloy Actuators: Design, Fabrication, and Experimental Evaluation, First Edition. Mohammad H. Elahinia.
© 2016 John Wiley & Sons, Ltd. Published 2016 by John Wiley & Sons, Ltd.

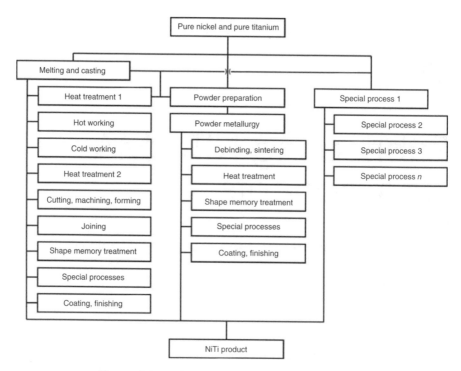

Figure 6.1   Fabrication routes for NiTi products

Figure 6.1 shows an overview of processes described in this chapter including a schematic of common manufacturing routes starting from elemental nickel and titanium toward the manufacture of the final NiTi product.

6.1 Melting and Casting

Since the functional and structural properties of NiTi strongly depend on the nickel–titanium ratio and are strongly affected by the impurity content, as was shown in Chapter 1, the melting and casting procedures are crucial steps in processing NiTi which require substantial precision and accuracy. In general, NiTi melts are highly reactive due to the high titanium content. As a matter of fact, high-temperature processing of NiTi, such as melting and alloying, is always accompanied by an increase in the impurity level. The pickup of impurities like carbon and oxygen results in the formation of Ti-rich phases, since the B2 phase in NiTi has a low solubility for both of these elements [1]. Carbon forms carbides of type TiC [2–5], while oxygen is dissolved in the Ti_2Ni phase and forms a stable phase of type $Ti_4Ni_2O_X$ [1, 6, 7]. The formation of these Ti-rich phases not only affects the structural properties by embrittling the material, it also degrades the functional properties and cyclic stability of shape memory or superelastic behavior. In addition, the formation of such phases leads to depletion of titanium in the B2 phase which causes a decrease of the phase transformation temperatures. In contrast, evaporation during melting and casting might contribute to a shift of the chemical composition in favor of the titanium content because nickel has a lower evaporation temperature than titanium.

Consequently, only highest purity raw materials as well as carefully selected refractory materials (e.g., crucibles and molds) have to be used. A precise control of the melting procedure is required to create the alloy with lowest impurity levels and the desired chemical composition. Taking into account the high reactivity of the melt, NiTi always has to be produced from elemental nickel and titanium by melting in an inert gas atmosphere or under vacuum. Vacuum arc melting (VAM) and vacuum induction melting (VIM) are the usual routes. Similarly, arc melting or induction melting is performed under a pure argon atmosphere. Melting under a moderate argon overpressure (0.4–0.6 bar) may even be advantageous over melting under vacuum because by applying pressure the vapor temperatures of the low boiling nickel are increased and evaporation of nickel is reduced [8]. It is worth noting that the setup and procedure are in principle very similar for melting in vacuum and under argon. Therefore, we do not differentiate between both methodologies and use instead the common terms VIM and VAM for these methods.

Both VIM and VAM are commercially used to produce NiTi ingots ranging from a few grams to several thousand kilograms [9, 10]. In terms of homogeneity, the VIM process is advantageous over VAM. In VIM, once the constituents are in the liquid state, electrodynamic stirring and mixing due to alternating current induction creates a high chemical homogeneity throughout the entire ingot [11]. The melting point of binary NiTi is about 1310°C so that VIM can be carried out at moderate temperatures. Temperatures between 1400°C [8] and 1450°C [12] up to 1550°C [10] are typical. Recommended crucible materials for VIM are calcia [12, 13], copper [8], and graphite, in particular [8–14]. Since the very first reports on NiTi melting procedures, it was clear that other types of crucible materials such as thoria [15] are not applicable due to their sensitivity to thermal cracking [10]. The major drawback of thoria and alumina- or magnesia-based crucibles, however, is their lack of thermodynamic stability. These materials are known to release oxygen, which can significantly increase the impurity of the alloy [10, 13]. Using graphite crucibles, oxygen contaminants are almost negligible, but a pickup of carbon is hardly avoidable. Hence, carbon levels of conventional VIM ingots are typically about 200–500 ppm [13] but can reach even 700 ppm [9]. The inevitable use of crucible materials, which easily react with the melt and lead to an impurity pickup, is a general disadvantage of VIM methods when compared to methods such as VAM that avoid the use of crucibles [9, 11]. There have been recent successful efforts toward lowering the carbon levels in VIM NiTi ingots by a very precise temperature control during melting and a special cladding method. Here, the crucible is charged in a way that a direct contact of the crucible and the melt is delayed as long as possible. This can be done by a cladding layer of titanium disks that act as a barrier in early melting stages between the crucible and the nickel, which melts first due to its lower melting temperatures [10]. Using this enhanced method, NiTi ingot with carbon levels of about 300 ppm can be produced by VIM [16].

As mentioned previously, VAM is advantageous over VIM in terms of chemical purity because there is usually no contamination from the crucible [8, 12, 13]. The main source of contamination in VAM is due to oxygen pickup from the furnace atmosphere which is comparable to the oxygen pickup during VIM. Nevertheless, VAM achieves the highest possible purity of the resultant alloy [8, 9, 13]. In the common VAM process, the raw material (nickel and titanium) acts as a consumable electrode which is melted by the arc and then solidifies directly in the water-cooled copper hearth or is dropped into a water-cooled copper mold where it solidifies. This procedure allows for the production of NiTi with carbon levels less than 200 ppm [12]. However, a single VAM step is generally not sufficient to produce a homogeneous ingot, because the melting process is limited to a small zone [12]. Only the upper portion of the volume is in the liquid state, while a small layer which is in direct contact with the water-cooled copper hearth remains solid. This makes it challenging to produce an ingot with a high chemical homogeneity. To promote thorough mixing, several additional melting steps are required [8, 12, 17]. These steps are then called arc remelting (VAR). Acceptable homogeneity will be reached after 5–6 remelting steps [8]. Similar to the zone melting or zone refining procedures, several VAR step also allow for reducing the impurity contents by localized melting. To keep a low impurity level, the ingot usually is machined after each remelting step to remove the oxide layer. This remelting method can also be applied to VIM ingots for further refining of VIM ingots. Today, combined melting processes, using VIM as primary melting followed by VAR remelt, are used in standard NiTi production. The final products are known as VIM/VAR ingots [8, 9, 12, 18].

Other melting methods besides VIM and VAR, such as cold skull melting and plasma or electron beam melting, are also in principle suitable for producing NiTi. These methods may even result in higher-quality NiTi alloys; they, however, are usually more challenging. Very low impurity contents for electron beam-melted NiTi have been reported [19]. The impurity is mainly due to the impurities of the raw material. This procedure, however, results in an unacceptable homogeneity of the alloy. Also, the evaporation that takes place due to the high temperature complicates the compositional control [8, 13, 19]. Additionally, processes other than VIM and VAR are hardly feasible from an economical point of view, and they are primarily used in laboratory scales [9, 12, 14].

Two novel and innovative melting approaches have been developed to manufacture porous SMA parts by liquid-phase processing. Young et al. [20] describe a method for NiTi-based SMAs (NiTiCu) where they use presintered SrF_2 salt foams as a space holder in a replica cast method. First, the alloy is melted under vacuum followed by applying gas pressure (argon, 1 atm) to infiltrate the SrF_2 foam with the liquid. After solidification and cooling, the space holder is removed by ultrasonicating. SMA foams with about 60% porosity that show a high deformation recovery were produced by this method. The resulting parts show changes in the transformation temperatures, which could take place due to the formation of secondary phases (oxygen-stabilized $Ti_2(Ni, Cu)$). Strictly speaking, the second method, described by Sugiyama et al. [21], is not a procedure to produce the alloy NiTi. This process is based on prealloyed NiTi materials. This method is a zone melting method under a hydrogen/helium atmosphere. While helium simply ensures a pure melting atmosphere, hydrogen is used as an injecting gas. Hydrogen is dissolved in the melt and is then rejected during directional solidification. This results in creating NiTi foams with aligned and elongated pores. It should be noted that hydrogen diffuses very rapidly in NiTi and its solubility in NiTi is in the order of over 40 at.-% which is more than enough to fully embrittle the material [22]. Even though these methods are attractive, they have not been commercialized.

Since conventional cast NiTi has inadequate structural properties and very low functional properties, usually thermal (homogenization and/or solution annealing) and multistage thermomechanical treatments (hot working, cold working including additional heat treatments) have to follow the melting process (exceptions exists for PM; see Section 6.5). These working procedures are described in the following section, and heat treatments are described in detail in Section 6.8.

6.2 Hot Working, Cold Working, and Forming

In conventional fabrication of NiTi products, forming procedures typically follow the melting step. Generally, NiTi ingots can be formed by the conventional hot and cold forming procedures. First, the ingot is usually forged, swaged, rolled, or extruded at elevated temperatures, mostly performed in successive steps with interpass annealing. Optimal hot working temperatures are about 800–950°C [12, 16] where the alloy is easily workable and oxidation is not severe [12]. These hot working steps reduce the dimension of the ingot while breaking down the cast structure, which is of great importance since the structural and functional properties of cast NiTi are not satisfactory.

The hot-worked material has a coarse microstructure due to dynamic recrystallization processes and grain growth. Such coarse grain structure is more prone to failure and less effective for functional properties. A reduction of grain size is beneficial for increasing mechanical properties such as fracture strength and yield strength. In addition, reduction of the grain size significantly affects the martensitic phase transformation and is therefore beneficial for functional properties such as shape memory and pseudoelastic properties as well as the functional stability of both effects [23–26]. A key issue behind this is that the stress required for either detwinning martensite (shape memory) or for inducing martensite (pseudoelasticity) has to be lower than the stress that causes dislocation slip. Otherwise, the material instead of showing the shape recovery behavior, as expected, will undergo large plastic deformation. Raising the stress for dislocation slip (yield strength) can be done effectively by grain size refinement and by introducing dislocations (work hardening) [27, 28]. Cold working (30–40%) is therefore an effective and common step in the processing route of NiTi.

Although NiTi is known to be workable in the cold state, actual cold working is substantially harder because the alloy work hardens quickly [12, 13]. In addition, shape changes due to stress-induced phase transformations as well as significant rebound or springback phenomena make precise cold working challenging. Therefore, it usually requires multiple cold forming steps with frequent interpass annealing treatments until the final dimension is obtained. Typical cold-working processes are rolling as well as wire and tube drawing. In the drawing process, several cross sections such as round or rectangular are common. Lubricants can improve the drawing process significantly [12] by reducing wear and breakage of dies or capstans and by cleaning the debris that remain in each of the drawing process [29].

It should be emphasized that the cold-worked NiTi does not exhibit the desired shape memory performance or pseudoelasticity yet. Cold work alone increases the strength of NiTi but leads to lower recoverable strains due to the introduction of random dislocations [27]. Proper functional properties are achieved by subsequent heat treatments [5, 8, 9, 12, 13, 16, 27, 28, 30]. For a description of these heat treatments, see Section 6.8.

Most of the forming processes are not suitable for manufacturing complex finished products made of NiTi. Wire drawing followed by bending processes such as spring coiling may be an exception for a forming process to prepare final or semifinal NiTi components (a heat treatment is still required). Another exception exists for braiding, which is used for stent manufacturing. This method rather is a joining than a forming technique and therefore not mentioned in this section (see Section 6.4). In general, forming processes are mainly used to produce semifinished NiTi products such as sheets, wires, rods, or tubes [31]. The choice of method depends on the required form of the semifinished part.

6.3 Machining and Cutting (Subtractive Manufacturing)

In general, machining of NiTi can be done by several conventional techniques such as (micro)milling, (micro)drilling, sawing, turning, shear cutting, and blanking. Material removing with cutting tools however is difficult because of massive burr formation due to the high ductility and elasticity of NiTi. Also, unfavorable chip breaking behavior, adhesion, work-hardening processes, stress-induced martensite, and springback effects make precise machining quite challenging. These effects usually cause considerable tool wear while degrading the quality of the workpiece [12, 32–34]. The machining techniques therefore have to be modified, and process parameters such as cutting speed and feed rate have to be adjusted for machining NiTi with acceptable results.

The most challenging machining operation is milling because the intermittency of the cut and the unfavorable material properties of NiTi lead to dulling or breaking of the tools. The process setup must be very rigid, and the tool must be extremely hard such as highest-quality carbide [33]. Despite these difficulties, there are approaches in the literature, which improve the effectiveness of milling NiTi by using special tools and adjusted process setup [35]. Turning NiTi on the other hand is not as challenging. Hard carbide tools and constant flooding with coolant are essential. In fact, turning was the method of manufacturing for producing the first commercial NiTi application for coupling hydraulic systems in the F14 jets [33].

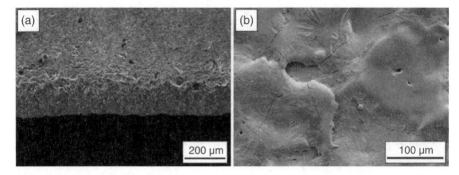

Figure 6.2   SEM micrographs showing a cut edge of a NiTi sheet machined by water-jet cutting with abrasive particles (a) [36] and the surface topography of NiTi after EDM process (b) [37]. Courtesy of Admedes Schüssler GmbH, Pforzheim (Germany). Reproduced with permission of Elsevier

In general, abrasive processes are more favorable for machining NiTi. Especially, various grinding processes, sawing, and erosive methods such as sandblasting work well [12]. According to Hodgson [33], problems can occur with a work-hardened surface layer resulting from grinding processes. In addition, these methods are usually not used for shaping but almost exclusively for producing semifinished parts or the final surface treatment of NiTi components. There may be few exceptions such as tips of guide wires which are tapered by grinding. In contrast to the aforementioned techniques, water-jet cutting and water-jet cutting with additional abrasive particles in particular (see Figure 6.2a) allow for machining final NiTi parts with intricate details. The achievable resolution of this technique is in the range of few hundred microns. Therefore, this method even allows for the fabrication of NiTi tubular stents [36].

NiTi can be processed by electro discharge machining (EDM) quite well [12, 38, 39]. The process variables such as pulse duration and discharge current, however, significantly affect the material removal rate (MRR) and the surface quality [40]. The rough surface after EDM either already demonstrates cracks (see Figure 6.2b) [37] or can introduce cracks which might decrease the fatigue life of a NiTi component [33]. The surface quality decreases with increasing the discharge current and pulse duration, and a high MRR results in the formation of an unwanted surface layer [41]. This surface layer usually contains impurities, mainly oxides, and also contaminants from the electrode [12, 39]. Also, a heat-affected zone (HAZ) of more than 100 μm depth may be present in the material [42]. Therefore, the surface structure and surface quality after EDM often is not acceptable for many applications, and additional finishing processes are required. Electrochemical treatments (see Section 6.9) however are well established and very appropriate to finish these surfaces [37]. With proper choice of process parameters, EDM can be used for machining intricate NiTi parts with a smooth surface and an average surface roughness of 1.2 μm [42].

Laser cutting is a key technology in machining NiTi. Thin sheets or tubes can be easily cut by lasers. Laser cutting is the most widely used method in manufacturing stents [12, 39, 43]. Especially for these intricate details, a high accuracy in the cutting process is mandatory. Today, cutting widths of less than 25 μm [33] and wall thicknesses of 250 μm [43] can be realized. To achieve this accuracy and the desired cut quality, it is necessary to have highly focused laser beam. In addition to the laser characteristics (type, wavelength, beam profile, and beam quality) and the process parameters (pulse width, power, feed rate, etc.), the ambient atmosphere affects the quality of parts. It is preferable to laser cut under a noble gas such as argon. The challenges associated with laser cutting include the occurrence of a HAZ, oxides, processing defects like inadequate surface quality, burrs at the cutting edges, and microcracks in the peripheral area which can propagate in the bulk material and might significantly reduce life time of the NiTi component. Postprocessing therefore is usually necessary for these parts to improve the expected life [12, 33, 39, 43, 44]. While pulsed Nd:YAG lasers are the most common, ultrashort pulsed laser like femtosecond lasers are more effective. With femtosecond lasers due to the short pulse duration, the material removal mechanism is similar to ablation. Thermal diffusion and conductivity effects are negligible, and the formation of a melted phase is suppressed; the material is rather sublimated. By optimized process parameters, the cutting process can be burr-free and does not create a HAZ (see Figure 6.3). The disadvantages of these lasers

Figure 6.3 SEM micrograph of high-precision laser-cut NiTi using ultrashort pulsed laser. Courtesy of Admedes Schüssler GmbH, Pforzheim (Germany)

are the high energy consumption, high prices, and comparatively low cutting speeds [43].

6.4 Joining

Joining of NiTi with similar or dissimilar materials (e.g., steel) and integrating NiTi components can in principle be carried out by different techniques. Simple methods include fastening and mechanical linkage or coupling techniques by interference fits. Press fits due to the shape memory effect can be easily used as shrink sleeves. Actually, the first commercial NiTi device was based on the use of this technique to press fit hydraulic lines, which were produced by machining. In 1969, Raychem Corporation (Menlo Park, CA) commercialized this tube coupling element (CryoFit™) for hydraulic systems for the F14 airplane produced by Grumman Aerospace Corporation (Bethpage, NY) [45]. Tight fits are usually not used to join NiTi components, but they are created by actuating NiTi components for locking and unlocking devices. An exception exists for tight fits realized by braiding. This technique has gained significant attention for manufacturing of stents (see Figure 6.4) [44, 46–48]. Clamping, clipping, and crimping are widespread joining techniques for simple NiTi components (e.g., wire products). Accidental releases under high mechanical load limit the transferable force of these joints. However, properly performed crimp connections can withstand several million load cycles [33]. Despite their simplicity, these joining methods may not be applicable in certain cases due to

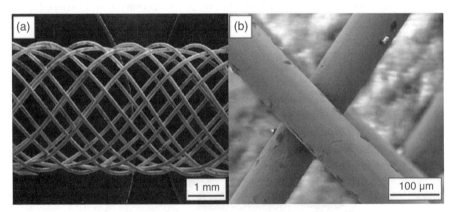

Figure 6.4 SEM micrographs of a braided NiTi stent: overview (a), higher magnification (b). Reproduced with permission from Ref. [44] of Wiley-VCH Verlag

geometrical restrictions. In these cases, and when higher loads have to be transferred, welding, soldering, brazing, and adhesive bonding are preferred.

For soldering NiTi to NiTi or to dissimilar materials (e.g., steel), usually low melting metals such as tin, lead, zinc, antimony, and bismuth are used, while copper, silver, zinc, indium, cadmium, etc. are applicable for brazing. In comparison to soldering, brazing allows for more stable joints which can bear higher loads. The disadvantage of brazing is the required temperatures (500–800°C) which are often higher than the temperatures, which are used in heat treatments for improving the functional properties (e.g., aging temperatures for improving pseudoelasticity). This, obviously, results in an inacceptable degradation of the NiTi component [33]. In addition, the mechanical properties of the soldered joints are not consistent and are usually well below those of the base material [49–51]. The same limitation applies to adhesive joints [8]. Also, low melting temperatures or softening at higher temperatures usually limits the use of the resulting devices in actuation. In addition, most adhesives are electrical insulators, which impedes the electrical activation of the resulting actuators [8]. Furthermore, toxic or hazardous constituents of many solders, fluxing agents, and adhesives can prevent the use of these joining techniques in certain applications such as surgical instruments or implants [33].

A novel brazing method for the manufacture of complex NiTi structures is based on the transient liquid reactive phase [52]. Binary NiTi and ternary NiTi-based corrugated sheets, discrete tubes, or wires are arranged to form cellular or honeycomb structures, wire space frames, sparse built-up structures, or discrete articles and then joined by transient liquid reactive phase brazing. As a braze material, pure niobium is used. Thereby, one makes use of the fact that pure niobium, brought into contact with NiTi, liquefies at a temperature which is below the melting point of both the brazing material and NiTi. The niobium then flows by capillary forces into the spaces between the NiTi elements and forms a strong joint. For this method, no flux is required which makes the technique even attractive for applications where biocompatibility is essential.

Generally, welding may be appropriate and preferable when a strong and durable joint of NiTi components is required. Although there are some approaches where resistance welding is successfully applied [53, 54], fusion welding processes (e.g., laser welding, electron beam welding, plasma welding, and arc welding) are the preferred methods. Due to the high reactivity of the alloy and its sensitive functional and structural properties, the fusion welding processes should be carried out under inert atmosphere or in vacuum [39]. Although the process control is more sophisticated, these fusion welding

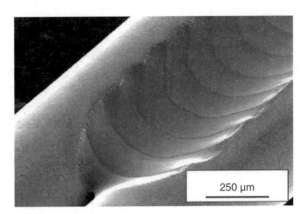

Figure 6.5   SEM micrograph showing detail of high-precision laser-welded NiTi wires. Courtesy of Admedes Schüssler GmbH, Pforzheim (Germany)

processes, and laser welding in particular, are widely used for various NiTi applications [39, 55]. In laser welding, CO_2 and TIG lasers are used; however, Nd:YAG lasers are the most common. The main challenges in achieving high-quality, defect-free laser welding of Ti-rich NiTi alloys are high tendency for hot cracking [56] and the formation of secondary phases in the HAZ and in the weld zone. These effects lead to a brittle material behavior which results in a significant degradation of the material properties [57–62]. These adverse effects can cause a reduction of both yield strength and elongation at fracture up to 50% compared to the base material. In addition, a shift of the phase transformation temperatures in the weld zone takes place as well as a change in the transformation behavior as welding tends to reset the effects of previous processing steps such as cold working and heat treatments [57, 60, 63]. This also results in a degradation of the functional properties in the form of lower reversible deformation and higher residual strains. In addition, evaporation of nickel in the weld zone can lead to a shift of the phase transformation temperatures, and the resulting depletion of nickel can also affect the pseudoelastic properties. Areas with lower nickel content transform at lower stresses into B19′ than the rest of the unaffected material [62].

Despite these challenges, laser welding of NiTi if performed properly in NiTi wires (see Figure 6.5), cross wires, or foils [54, 56, 64, 65] as well as in dissimilar NiTi joints (NiTi to other materials such as stainless steel) [56, 65] will result in satisfactory structural and functional properties. Subsequent heat treatments may be applied for further improving the functional properties [64]. For welding dissimilar joints, a process modification is preferable to avoid the formation of brittle intermetallic phases such as TiFe and Fe_2Ti when joining NiTi with steel. One possible solution is the application of an interlayer of pure tantalum, pure nickel, or pure cobalt as filler in the fusion zone [65–69]. Good results were also achieved without additional filler material but by simply displacing the melt zone from the center of the joint. By moving the laser to the NiTi side of the NiTi–steel joint, the chemical composition of the melt can be affected in a way to significantly reduce the formation of brittle intermetallic phases [56].

6.5 Powder Metallurgy

Generally, PM is highly attractive for processing NiTi, as it usually provides near-net-shape devices. This may overcome the challenges in conventional processing of NiTi and thus may contribute to a significant simplification of the fabrication process. In addition, some PM methods also allow for the production of porous NiTi which is very attractive for biocompatible implants and damping applications. Nevertheless, many PM routes are characterized by a significant drawback against conventional machining. Due to the large specific surface area of the powder particles and due to at least two additional high-temperature processing steps (powder production, PM process), the resulting NiTi components usually contain high contaminant levels [12]. Oxygen contents in NiTi PM devices may even be higher than 1500 ppm [39]. As described before, high impurity levels may considerably degrade structural and functional properties of NiTi. Nevertheless, there are several already established PM routes as well as some promising PM approaches for the production of NiTi which are described in this section.

Usually, PM processing methods use prealloyed NiTi powders. Some exceptions may exist for special methods where alloying NiTi from elemental nickel and titanium powders and shaping are combined in one processing step. Using such a mixture/feedstock of elemental powders, however, a precisely controlled process (e.g., sintering temperature) is crucial. Otherwise, due to different diffusion behavior and diffusion velocities, Ni- or Ti-rich precipitates may form in addition to NiTi. Also, areas may remain which still consist of pure nickel or titanium [70, 71].

The initial step for a PM manufacturing method of NiTi components is powder preparation. Similar to production of NiTi ingots, powder production of this material is challenging. In general, there are different ways for preparing powder either from NiTi or from elemental nickel and elemental titanium. These methods include mechanical attrition or ball milling, atomization in water or in gas, creating powders by laser ablation, and hydriding. Each method is characterized by specific powder properties such as homogeneity and chemical purity, particle size distribution, and particle shape. The size of the particles, their distribution, and their shape strongly determine the physical properties of the powder such as bulk density, compacted density, and flowability [72–74]. It is worth noting that powder preparation usually results in impurity pickups and the impurity content increases with decreasing particle size due to the higher surface to volume ratios. In hydriding, specifically, the hydrogen content in the powder may decrease the functional properties of the resulting parts. Mechanical powder preparation may introduce fewer impurities than hydriding or atomization from the melt, but mechanical preparation usually is quite challenging because of the poor workability of NiTi. The other benefit of mechanical preparation is that it allows for alloying NiTi powders from pure nickel and pure titanium without undergoing a liquid phase. The shape of mechanically alloyed powders usually is irregular. This also applies to water atomized powder due to the high solidification rates. Another drawback of water atomization may be the diffusion of hydrogen into the NiTi. For most powder metallurgical processes, therefore, gas atomization of NiTi seems to be the most applicable method [22]. The powder shape is almost spherical due to lower solidification rates compared to water atomization, and if performed in an inert atmosphere, the impurity levels can be kept comparatively low.

Different gas atomization methods such as the electrode inert gas atomization (EIGA) process or the NANOVAL process result in different powder characteristics. Both of these processes are especially feasible for atomization of reactive and sensitive materials. In the NANOVAL process, the alloy is melted in a graphite crucible and flows through a nozzle at the bottom of the crucible. A gas (typically argon) flows from around the melt nozzle into a Laval nozzle. The melt monofilament becomes attenuated (thinner and thinner) until it bursts open spontaneously [75]. This method produces fine and ultrafine powder of spherical shape in narrow particle size distribution. The drawback is that the melting takes place in a crucible, which results in impurity pickup [76]. In contrast to this method, the EIGA process does not use a crucible. Instead, the ingot is melted contact-free in an induction coil. The melted alloy is then dripped from the ingot through a gas nozzle system (atomizer), where it is atomized by argon (see Figure 6.6a) [77]. Because of the crucible-free melting process, this method allows for production of a NiTi powder with very low impurity contents [74, 76, 78]. The drawback of this method is a wide range of particle sizes in the resulting powder [76]. This can be seen in the SEM picture in Figure 6.6c. There are a significant number

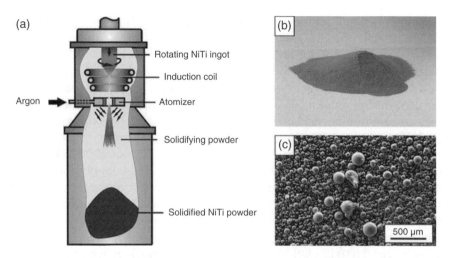

Figure 6.6 NiTi powder preparation using the EIGA method: schematic of the EIGA process (a), atomized NiTi powder (b), SEM micrograph of the NiTi powder (c). From Ref. [74]

of larger nonspherical particles with irregular shapes. A sieving and screening procedure is therefore usually required before further processing the EIGA powders in any PM method.

The simplest PM method is the conventional sintering of loose or compacted powders at a temperature close to the melting point of NiTi. In this common production method for NiTi parts, both mixtures of elemental powders [79–84] and prealloyed NiTi powder [85] are used. For sintering elemental powders, it is possible to use TiH_2 powder instead of pure titanium or in addition to the pure titanium powder. In this process, the hydride acts as a blowing agent which decomposes and releases hydrogen. The addition of TiH_2 accelerates the reactive sintering process and decreases oxidation of the titanium. There might be the risk of embrittlement when the hydrogen is solved in NiTi. This processing route however is quite efficient because TiH_2 powders are less expensive than pure titanium [79]. When processing elemental powders (or TiH_2), however, proper process control and the adjustment of adequate sintering parameters are essential for producing single-phase NiTi. In the case of using inappropriate process control and sintering parameters, insufficient diffusion processes may result in pure nickel or pure titanium as well as secondary phases [79, 80, 86].

Since conventional sintering has a limitation in part geometry, this method does not offer potential for fabrication of complex NiTi components. Instead, sintering often is used for producing semifinished NiTi parts. Still, sintering can be attractive for processing porous NiTi or NiTi foams. Conventional sintering does not allow for an independent control of pore size. The result is parts with low porosities and small pores, which are dictated by the initial powder size. To increase porosity and the pore size, temporary space holders are used during sintering [87–89]. In this process, a mixture of elemental powders and ammonium acid carbonate powders (NH_4HCO_3) is cold compacted and then heated up to moderate temperature to remove the space holder. Finally, these green parts are sintered to produce highly porous NiTi with porosities up to 87% and pore sizes of up to 500 μm [88]. These highly porous samples however do not exhibit sufficient mechanical and structural strength. In a similar transient liquid-phase sintering approach, NaCl particles are added as temporary space holders to a blend of NiTi and pure niobium powders (note: additional pure nickel powder can also be added to the mixture for alloying more Ni-rich NiTi) [86, 90]. Due to the addition of niobium, which is acting as a transient liquid-phase sintering agent, a eutectic reaction at 1170°C, which is 140°C below the melting point of NiTi, takes place in the sintering process.

Spark plasma sintering (SPS, also known as field-activated pressure-assisted synthesis) [71, 91, 92] as another sintering method is used for production of dense NiTi and porous NiTi. In SPS, a feedstock consisting of either a mixture of elemental powder [92] or prealloyed powder [91] is compacted usually in a graphite or steel die [92]. This compacted feedstock is then exposed to a high electric current density. Joule heating causes diffusion, which initially leads to densification and finally to homogenization of NiTi. One of the benefits of SPS is its very short reaction time and low temperatures. This may avoid high impurity pickups [91], but secondary phases may still be present [92].

A method to create interconnected elongated pore channels or networks in hot pressing of NiTi (see Figure 6.7) consists of using low-carbon steel wires which are woven into orthogonal meshes as space holders [93]. These meshes are pack carburized in pure carbon powders to supply carbon to the wire surface so that a TiC layer can form at the steel–NiTi interface which prevents interdiffusion between the two phases during the later hot-pressing procedure. These meshes and prealloyed NiTi powders are then poured into a die in an alternating manner and hot pressed under vacuum. After densification and cooling, the steel wires are removed electrochemically. Using this method, interconnecting and accurately shaped pores with tailored design and arrangement can be fabricated. There however remains a layer of TiC on the surface of the channels which can affect the structural and functional properties of NiTi.

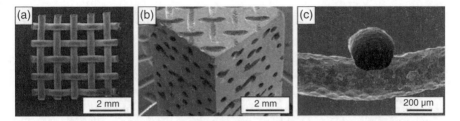

Figure 6.7 SEM micrographs of a steel mesh space holder (a), of a porous NiTi scaffold produced by hot pressing using these space holder (b) of the pores at higher magnification (c). Reproduced with permission from Ref. [93], Elsevier

In synthesis processes such as explosive shock-wave compression or self-propagating high-temperature synthesis (SHS, also known as combustion synthesis), elemental powders are used, exclusively. The success of these processes is due to the fact that the synthesis of pure nickel and titanium to NiTi is an exothermic reaction type:

$$Ni + Ti \rightarrow NiTi + 67\,kJmol^{-1}\,[82]$$

By means of an initial reaction, locally a melt pool evokes where the phase NiTi is produced and, through a self-sustaining reaction that propagates through the powder, the NiTi phase is finally formed in the entire volume. NiTi components produced in this way are usually highly porous with interconnecting pores [94]. However, SHS results in uncontrollable and inhomogeneous pore sizes and chemical inhomogeneities are also difficult to avoid [95]. In particular, undesirable intermetallic phases of type Ti_2Ni and Ni_3Ti are usually formed in accompanying reactions [39, 96, 97]. This may be due to the temperature gradients during processing. It is worth noting that the maximum temperature in the reaction zone can be lower, equal, or slightly higher than the melting point of NiTi [94]. Nevertheless, a porous NiTi implant for intervertebral fusion was successfully produced by SHS and is commercially available by Biorthex (Montreal, Canada) under the product names Actiporetm PLFx and PNT (see Figure 6.8) [98, 99].

For producing almost dense binary NiTi SMAs or NiTi-based ternary SMAs from loose, prealloyed, or elemental powders, hot isostatic pressing (HIP) is appropriate [100–102]. The difficulties that may be encountered with the use of elemental powders to fabricate NiTi via HIP are similar to those encountered using conventional sintering like the formation of unwanted secondary phases ($NiTi_2$ and Ni_3Ti) as well as pure nickel and pure titanium and

Figure 6.8 Porous NiTi interbody fusion devise ActiporeTM produced by SHS: (a and b) photography and (c), SEM micrograph. Reproduced with permission from (a) Ref. [103], (b) Ref. [104], and (c) Ref. [105], John Wiley & Sons, Ltd

deviations in the chemical composition in the NiTi phase [100, 106]. Due to the difficulties in control of the exothermic reaction and the Kirkendall effect, HIP of elemental powders may not be effective [76], but the reaction may be more stable and controllable compared to SHS [106]. The HIP-produced NiTi from prealloyed powders has a significantly more homogeneous microstructure and shows significantly better mechanical properties than HIP-produced NiTi from elemental powders [100]. In mechanical testing, the material produced from elemental powders does not even reach the plateau stresses for stress-induced transformation. In HIP, the powder is usually encapsulated under vacuum or under noble gas atmosphere in a container before being sintered. A capsule-free HIP production has also been developed [107, 108]. In both of these HIP routes, simultaneous application of isostatic pressure and temperature leads to a uniform density in the sintered part [73, 101, 109, 110]. HIP may be advantageous over the other PM methods because of a decreased solid-state diffusion time [106]. In addition, being protected from atmosphere due to the use of the capsule may limit the impurities pickup during HIP.

Porous NiTi has successfully been made with various HIP-based methods [95, 106, 108, 111–116]. The resulting pore size is relatively small of about 20 µm [95] in conventional HIP. By slightly modifying the HIP procedure, however, highly porous NiTi with large pore sizes can be produced [95, 111]. In this approach, the expansion of the noble gas (argon) entrapped in pores is used to increase the pore size to approximately 500 µm [95]. The approach includes a partial liquid-phase sintering stage to enhance the homogeneity of the alloy, but some Ni-rich phases are still observed in the material. Temporary space holders are also added [95, 107, 115, 116] to elemental powder mixtures [107, 115] or to prealloyed NiTi powder [95, 116] for producing porous NiTi via HIP. These space holders can be NH_4HCO_3 particles, for example [107, 115], NaF [95] and NaCl [116] however may be more preferable because of

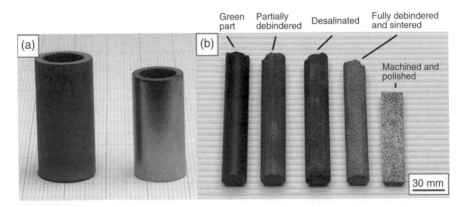

Figure 6.9 NiTi MIM parts: shrink sleeves directly after MIM processing (left) and after subsequent debinding and sintering (right) (a), figure courtesy of Dr. Martin Bram, Forschungszentrum Jülich GmbH, Jülich, Germany. Porous NiTi samples produced via MIM using space-holders at different stages in processing (b). Reproduced with permission from Ref. [109], Elsevier. Reproduced with permission from Ref. [117], Elsevier

their thermodynamic stability in contact with nickel and titanium. In the first approach, the feedstock is cold compacted and the NH_4HCO_3 space holders are subsequently removed by heating before performing the HIP procedure [107, 115]. In the second method, the space holders are removed after the HIP procedure [95, 116]. The disadvantage of removing the space holder before HIP is in increasing the possibility of pore collapse. In addition, the use of NaF or NaCl space holders allows for an easy removal because they can be dissolved in water. The second approach has shown comparatively high mechanical and structural characteristics in the resulting porous NiTi [116].

While HIP enables the net-shaped production of parts with only moderate geometrical complexity (e.g., tubes or cylinders) [102], metal injection molding (MIM) offers near-net-shape fabrication of more complex parts. MIM is also very promising for the production of dense and porous NiTi (see Figure 6.9) [78]. Part sizes from 2 to 50 mm can be processed with masses of up to 1 kg [76] while maintaining a high throughput, which makes this method efficient for manufacturing. In MIM, NiTi powders or elemental nickel and titanium powders and a binder system (usually amide wax or polyethylene wax) are blended by means of mixing in a heated kneader. This homogeneous mixture is then injected in a heated MIM tool. After part removal and cooling, a debinding step at elevated temperature is carried out to remove the binder system, and a subsequent sintering procedure of this green body is performed. For NiTi processed in this manner from prealloyed powders, very good functional and structural properties are achieved [78, 118, 119]. Therefore, most approaches exclusively use prealloyed NiTi powders [76, 78, 117, 120–123], but it is also shown that elemental nickel and titanium powders can be processed via MIM [124]. For producing highly porosity NiTi with large pore sizes, the space-holder technique (using NaCl) is adopted for MIM [76, 78, 117, 120–122]. In this procedure, the MIM processing route is extended by an additional step to remove the space holder (see Figure 6.9b). When using NaCl, the space holders can be easily dissolved in distilled water before the final sintering step is carried out. The resulting material has promising structural and functional properties [117, 120] with high porosity (~50%). The parts show up to 6% pseudoelasticity under static compression loads.

Over the last 10 years, additive manufacturing (AM) has gained significant attraction for processing NiTi [74, 125–152]. Generally, the term AM describes processes which are used to create physical parts directly from CAD data by adding material in sequential layers. For AM of metals, these layers are usually provided in powders and melted by a laser. It is worth noting that although other energy sources such as electron beam are also available, for processing NiTi, only lasers are used. AM facilitates fabrication of highly complex parts, which cannot be processed by any other method [153]. This fabrication method therefore provides a freeform of fabrication principle, which circumvents current manufacturing challenges. Lattice-like and truss-based structures, curved holes, designed porosity, hollow parts, and other intricate features are realizable. The most common AM techniques for making parts from NiTi are powder-bed-based technologies like Selective Laser Sintering (SLS), Selective Laser Melting (SLM), Direct Metal Laser Sintering (DMLS), and LaserCusing. Figure 6.10 describes the main steps in a powder-bed manufacturing. As the first step, a 3D-CAD model of the part is sliced in horizontal layers. Each layer contains specific information about the part geometry and the scan trajectories for the laser. As Figure 6.10 shows, AM is an iterating process. According to the thickness of the sliced CAD layers (typically 20–100 μm), a blade, knife, or a roller deposits a powder layer with the desired thickness. The laser beam locally melts the powder according to the geometry of the part. After solidification, solid structures remain which are surrounded by loose powder. Afterward, the next powder layer is deposited on top of the previous layer. This procedure is repeated until the desired 3D part is produced.

Other AM techniques that are used to process NiTi are flow-based methods, like laser engineered net shaping (LENS) [130, 134] and direct metal

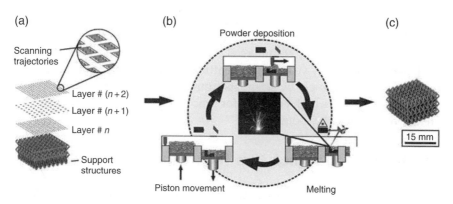

Figure 6.10 Schematic showing the principle of powder-bed-based additive manufacturing of complex NiTi scaffolds: CAD data preparation (a), cyclic AM procedure (b), additively manufactured NiTi scaffold (c)

deposition (DMD) [126, 139] which are very similar to the conventional deposition welding. Here, the powder is deposited through a nozzle and not as layers. The nozzle directly feeds the powder into the laser focus. Powder-bed-based technologies are, however, used most often for NiTi as they allow for more complexity in the produced parts.

For processing NiTi, the entire AM process is usually done in an enclosed chamber filled with argon to minimize oxidation. While elemental powders can be used [125, 128, 131, 139, 150], prealloyed powders are more popular [74, 125–127, 129, 130, 132–138, 140–152]. By using elemental powder mixtures, similar difficulties occur as described in other PM processes. Other intermetallic phases as well as pure nickel and pure titanium may be present in the resulting part. These issues are generally resolved by using prealloyed powders [145]. Using prealloyed powders, the additively manufactured material is quite homogeneous and shows very high strength and functional properties which are quite similar with the properties of conventionally processed NiTi [74, 140, 146, 147]. Ti-rich NiTi exhibits the shape memory effect directly after processing [74, 140, 146], while Ni-rich NiTi requires subsequent heat treatments to induce pseudoelastic properties [74, 143]. Furthermore, AM NiTi parts can have very low impurity contents which makes AM attractive for manufacturing of medical devices. AM NiTi parts can meet the impurity limits for medical NiTi, prescribed by ASTM F2063-05 [74, 133, 146]. In addition, AM offers the advantage of creating engineered porosity by the CAD design. This opens up the possibility to design interconnected pores, variable pore sizes, or morphologies and complex scaffolds that can be stiffness tailored. AM therefore is very attractive for producing patient-customized implants from data derived by imaging methods such as CT scans. Studies on the biocompatibility of AM NiTi have shown promising results [137, 141, 149].

AM of NiTi has not been commercialized. Metal processing AM systems are available since the early years of this century, and these techniques were used for processing NiTi solely by scientists from academia for less than 10 years. This method provides high potential for processing NiTi, and the industry has already recognized this potential.

6.6 Thin Film and Thick Film Technologies

NiTi thin films with thicknesses ranging from less than 15 μm [154] to 100 μm have attracted much attention as miniaturized high-performance sensors or actuators in microelectromechanical systems (MEMS) [155, 156]. Due to the small mass of NiTi thin films, the cyclic (heating–cooling) response time can be reduced substantially, and the speed of operation may therefore be increased significantly. In addition, the work output per unit volume of NiTi thin film exceeds that of other microactuation mechanisms [157, 158]. The first studies on producing NiTi thin films were published in the 1980s [159, 160]. These thin films are deployed in various actuation mechanisms such as miniaturized wrappers (see Figure 6.11) and miniaturized pneumatic valves and for haptic displays or as spacers in flat panel displays [12]. Also, biomedical applications such as transcatheter heart valves are based on the properties of NiTi thin films [154].

NiTi thin films are usually produced by deposition methods. The most commonly method is (magnetron) sputtering [13, 154, 155, 158, 162]. Other

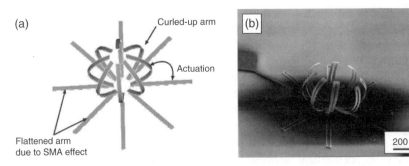

Figure 6.11 NiTi microwrapper fabricated by sputtering NiTi on a patterned substrate: schematic of activation principle (a), SEM micrograph of the wrapper (b). Reproduced with permission from Ref. [161], Elsevier

methods such as laser ablation, ion beam deposition, arc plasma ion plating, plasma spray, and flash evaporation are also reported. The intrinsic problems with these methods include nonuniformity in film thickness and composition, low deposition rate, and uneconomic processing [13, 158]. Despite its popularity, the sputtering process is challenging because sputtering parameters and conditions such as target quality, target power, gas pressure, deposition temperature, substrate, and target position and distance have to be optimized to achieve and control the desired chemical composition and to guarantee proper functional and structural properties such as the exact desired phase transformation temperatures. Also, the choice of the proper substrate material is essential. The substrate material will considerably affect stress generation and stress evolution in the film if both materials have significant mismatches and if there are differences in their thermal expansion behavior [158]. Most common substrate materials for NiTi films are rock salt [13], (silicon-based) glasses, or wafers of pure silicon [158, 160, 163–167]. Also, deposition on a polyimide substrate is possible [162]. The use of elemental nickel and titanium targets is difficult as the sputtering yield of nickel is higher than of titanium, and hence, the stoichiometry of the deposited film tends to be titanium deficient. In addition to the mismatch in the sputtering rates, there is also a steady change of the rates with overall sputtering time. Several approaches are developed to overcome these obstacles. For example, the use of prealloyed targets can achieve and maintain the desired film stoichiometry. In addition, using prealloyed targets results in more uniform compositions across the film [12]. However, even using prealloyed NiTi targets, sputtering involves a significant loss of titanium. One can circumvent the titanium depletion by placing additional titanium on top of the alloyed target [168, 169] or by using targets which have higher titanium contents than the desired composition of the resulting film [158, 170–174]. Increasing the NiTi target temperature can also produce a compositional modification by limiting the titanium loss [158, 163, 164].

It is difficult to sputter three-dimensional objects. One way to achieve these structures is to use conventional sputtering on planar substrates in combination with photolithography [175]. In this technique, a three-dimensional NiTi thin film structure is produced by depositing multiple layers while sequentially interspersing layers of a sacrificial material such as chromium. These chromium layers, which are partially removed in the subsequent chemical etching processes, form bonds between the NiTi layers at selected locations and serve as a mask for structuring and patterning the NiTi layers in the later photolithographic process.

Generally, NiTi films deposited at ambient temperatures are amorphous in the as-deposited state and require subsequent heat treatments for recrystallization. These heat treatments have to be carried out in a vacuum to limit oxidation and at temperatures higher than 450–500°C [158, 165]. Typical recrystallization treatments for NiTi thin films are 600–650°C for 1 h [170, 171]. Crystalline films in the as-deposited state can be achieved by sputtering at elevated temperatures or using heated substrates [12, 158] as films deposited at temperatures about 400–500°C crystallize in situ. During sputtering, the initial temperature can be decreased to approximately 300°C which is sufficient to maintain a crystalline growth during the later sputtering process [158].

Thick film techniques are commonly used to apply NiTi layers or coatings at thickness of tens to hundred of microns. These coated parts benefit from the superior properties in reducing corrosion [176–179] and/or wear, erosion–corrosion, and cavitation of NiTi [180–184]. In special cases, thick film technologies are also used to produce stand-alone NiTi semifinished parts such as thin-walled foils [185]. Thick film techniques include thermal spraying via vacuum plasma spraying (VPS), atmospheric plasma spraying (APS), high velocity oxygen fuel spraying (HVOF), or laser cladding techniques. These processes are based on using mostly prealloyed NiTi and in certain cases elemental powders [181]. In addition to powder processing spraying techniques, (low pressure) wire arc spraying (WAS) is also reported in which NiTi wires are melted by an electric arc and then sprayed through a nozzle [185].

6.7 Heat Treatments and Shape Setting

NiTi heat treatments including thermomechanical treatments are complex procedures that significantly affect structural and functional properties of the alloy. While these treatments appear to be quite simple and straightforward at a first glance, the reality is that there exist several complexities and interdependencies [33]. These processes usually have to be developed or optimized individually depending on the processing history and the resulting microstructure, objectives of the heat treatment, product type, and application type or desired properties. These requirements in some cases result in processing conflicts. A brief overview of the most important treatments and some general characteristics are described here. For further reading, the relevant and comprehensive literature edited and published by Otsuka, Wayman, Ren [155, 186], and Funakubo [187] are recommended.

Directly after melting, NiTi usually has to be heat treated to eliminate casting and solidification artifacts such as segregation, microscale concentration gradients, and precipitates. Homogenization treatments follow the alloying or an initial powder metallurgical processing step. While some of these treatments are performed at temperatures of 1000°C [16, 188–190] and higher, it is desirable to keep the temperature below 984°C to avoid melting of the Ti_2Ni phase which easily forms in Ti-rich NiTi. This phase is stabilized by oxygen and could alter the microstructure and affect the transformation temperatures. Typical parameters for homogenization treatments of NiTi are 950°C for 24 h under vacuum followed by quenching [17].

Some high-temperature processing routes, such as PM, require intermediate solution annealing treatments because during processing at elevated temperatures, precipitates may form in Ni-rich NiTi alloys. Temperatures for this treatment are similar or lower than the temperatures required for homogenization [12, 191–193]. The durations for solution annealing are usually shorter than homogenization but may differ in a wide range from 5 min to 5.5 h depending on the processing route.

Cold forming of NiTi performed in successive steps requires intermediate heat treatments to restore the workability. NiTi shows significant work-hardening effects. Therefore, it is essential to anneal and recrystallize the alloy frequently to create a microstructure which can tolerate further deformation without fracture. These heat treatments are effectively performed above 600°C [33]. The most common treatments are in the range of 700–900°C for a few minutes (1 min to 0.5 h) [16, 27, 33].

In addition to these intermediate heat treatments, NiTi products (bars, wires, ribbons, and sheets) which received a final cold-working step require further heat treatment to (i) eliminate the distortion which usually is a forming artifact in a straight annealing process [16, 30], for example, for wires after cold drawing and to (ii) establish the desired structural and functional properties in the alloy since cold working suppresses the shape memory response and pseudoelasticity as randomly introduced dislocations hinder the mobility of the twin boundaries. In some cases, this heat treatment can be combined with the shape setting treatment (see below) or with sophisticated electropulsing heat treatments, for example, in cold-drawn wires [194]. However, for achieving optimized properties, materials with 30–40% retained cold work should be heat treated in the range of 350–500°C. For shape memory materials, this treatment is usually performed at temperatures between 350 and 400°C, while Ni-rich pseudoelastic materials require treatments with slightly higher temperatures up to 500°C [12]. This difference is attributed to differences in microstructural mechanisms for both materials. For pseudoelasticity, the presence of Ni_4Ti_3 precipitates which are formed during this aging treatment [188] is extremely beneficial. In Ni-rich alloys, the formation of such Ni-rich phases provides a precipitation hardening of the material and therefore leads to an increase in yield strength. Consequently, when the material is loaded to stress levels where transformation into martensite occurs, it receives less dislocation slip and will therefore show less irreversible strains after unloading. In addition, Ni_4Ti_3 precipitates basically represent nucleation sites, and their presence enhances the martensitic transformation because the required critical stresses are reduced. Both effects promoted by aging and precipitation of Ni_4Ti_3 particles are essential for proper shape recovery in pseudoelastic operations [1, 155, 189, 190, 195–201].

It is important to note that aging treatments for Ni-rich alloys, which result in precipitation, affect the transformation behavior as well as the mechanical behavior of the material. While a material which is free from precipitates or has a low precipitate density behaves more ductile and shows large strains until crack initiation, crack growth, and failure, the precipitation hardening effect in aged materials results in a more brittleness [195, 201, 202]. In addition, the phase transformation in Ni-rich alloys is significantly affected by aging treatments. By formation of Ni-rich precipitates, there is corresponding nickel depletion in the matrix, which leads to an increase in the transformation temperatures [1, 191, 198, 203]. Also, the presence of Ni_4Ti_3 precipitates creates microscale stresses in the material which can alter the transformation sequence. Instead of a single-step transformation between austenite and martensite, a two-stage or a multistage transformation undergoing the R-phase is observed [191, 199, 203–209].

Recently, novel annealing methods are investigated to fabricate NiTi parts with locally tuned transformation behavior [210–216]. Using this method, the microstructure of a wire, rod, sheet, or thin film is only locally annealed, crystallized, or aged most commonly by a laser. Local annealing in a furnace is also reported [215]. As a result, the device processed by this method exhibits different actuation temperatures and therefore shows multiple memory functions (see Figure 6.12). Even the combination of shape memory and pseudoelasticity in a single device can be established [216].

As indicated in the left part of Figure 6.12, semifinished SMAs such as wires require a shape set configuration to reveal the demanded functional properties before being used in their desired application. This configuration is accomplished by a special heat treatment called shape setting. The objective of this treatment is to improve the shape recovery behavior since cold-worked NiTi does not exhibit shape memory properties and to establish the geometrical shape, which is memorized in the shape memory operations. For this heat

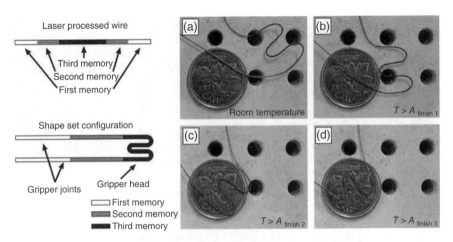

Figure 6.12 Locally annealed NiTi microgripper: schematic of annealing procedure and shape set configuration (left), microgripper during sequential activation of each embedded shape memory response (right). Labels (a–d) shows the progression of the actuation of the system. Reproduced with permission from Ref. [214], SAGE Publications

treatment, the semifinished product such as a wire or a sheet is cold formed on a jig into the final product shape (see Figure 6.13). This cold deformation requires significant amount of deformation beyond the recoverable limit because of the large springback effect in NiTi or pseudoelastic shape recovery. In this constrained shape, the NiTi has to be fixed tightly on the jig, for example, with screws or by using a tight fit. The suppression of free recovery results in the buildup of stresses within the material, which later will relax in the course of the following shape setting heat treatment [16]. If the material is not restrained during this heat treatment, the shape will revert partly to the initial configuration [12, 33].

The shape setting heat treatment can be performed at moderate conditions, but the temperature has to be about 350–450°C higher than the austenite-finish temperature of the alloy [33]. Temperature and duration may vary slightly to account for the desired application and geometry (wire, strip, ribbon, sheet, etc.), but both significantly affect the shape recovery properties, mechanical behavior, and transformation temperatures of the material [218]. As shown in Figure 6.14, shape setting at low temperatures results in low recoverable strains, but if the deformation is suppressed, high forces can be achieved because work-hardening effects are mostly retained. In addition, shape setting at low temperatures also results in higher fatigue life.

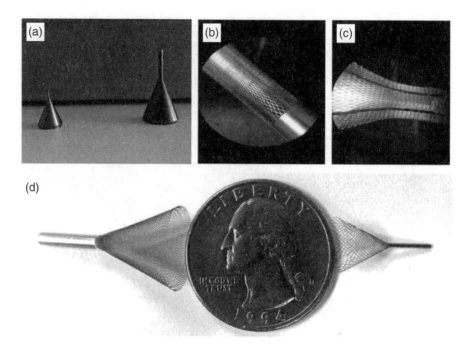

Figure 6.13 Steps of shape setting procedure for a minimally invasive thrombectomy device: shape setting jigs (a), laser-cut stent-like structure before being shape set (b), expanded structure after shape setting (c), final thrombectomy device proximal funnel and distal basket relative size to a quarter (d). From Ref. [217]

In contrast, shape setting at higher temperatures results in higher recoverable strains, but achievable forces and fatigue life decrease [8]. It is noted that the plot in Figure 6.14 refers to an annealing duration of 30 min. If shape setting is performed for varying durations, the general tendencies will still be valid qualitatively, but the value for maximum stresses and recoverable strains may differ. Typical shape setting parameters for NiTi wires for actuation applications are 400°C for 5 min followed by quenching [16]. For this heat treatment, fluidized beds are well established because this type of furnace guarantees a very uniform temperature profile and the thermal mass and conductivity of the particles in the fluidized bed leads to rapid heating of the part to the desired temperature. In addition, the inert particles (e.g., alumina) and the flowing noble gas (e.g., argon) prevent significant oxidation [33]. The shape setting process is usually performed only for single parts or small

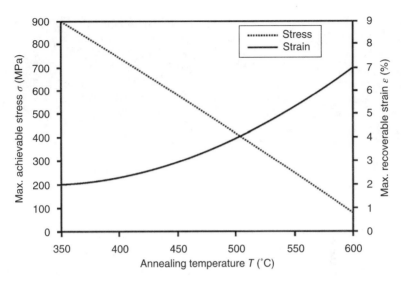

Figure 6.14 Schematic showing the correlation of annealing temperature for shape setting (t = 30 min) of a NiTi actuator and maximum achievable stress of this actuator in case of suppressed deformation and maximum recoverable strain in case of free shape recovery, respectively. From Ref. [8]

batches. However, this process can be scaled up to production quantities by increasing the number of jigs and increasing furnace capacities, and for some NiTi products, it can be semiautomatized by using forming machines such as coiling machines for producing shape memory springs [12].

6.8 Finishing and Surface Modification

Several NiTi applications require surface finishing for various reasons such as cleaning, smoothening, structuring, or modification. Most NiTi semifinished parts (e.g., sheets, wires, bulk, etc.) require a cleaning step to remove oxide coatings, contaminants, or lubricants which remain on the surface after processing [33]. Washing with detergents is effective for removing most lubricants, but caution should be used if using alkaline cleaners which can lead to embrittlement and pitting [33]. NiTi processed at high temperatures typically has significant oxide surface layers. Also, NiTi which received a heat treatment (in air) above 300°C shows considerable surface layers consisting of oxides of type TiO_2 and intermetallic phases of type Ni_3Ti [8]. These dark oxide layers can be effectively removed mechanically, by grit blasting,

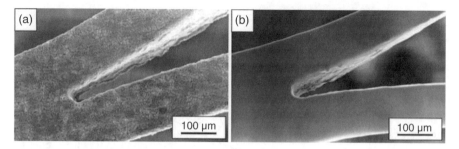

Figure 6.15 SEM micrographs showing the surface of a NiTi stent before (a) and after electropolishing (b). From Ref. [222]

polishing, or tumbling with fine abrasives. By proper selection of polishing media, a mirrorlike finish can be achieved by mechanical polishing [12, 33].

Most acids do not significantly affect the inert, adherent TiO_2 layer. It is possible, however, with extremely hazardous acids such as hydrofluoric acids to remove the surfaces oxides. In addition to being hazardous to environment and human health, these chemicals can cause hydrogen pick up which can lead to hydrogen embrittlement of the exposed NiTi parts [33].

Electropolishing is a very appropriate method for optimizing NiTi surfaces especially for biomedical applications [8, 12, 33, 219–221]. Most established electrolytes for NiTi are mainly based on perchloric acid, acetic acid, and phosphoric acid [8, 219, 222]. In particular, stents (see Figure 6.15) and other biomedical NiTi components produced by laser cutting can be post processed by electropolishing. This process not only removes the oxide surface layer and the HAZ, it also smoothes the surfaces and blunts sharp edges which remain after laser cutting. Electropolishing is therefore helpful in minimizing the potential trauma to the vascular tissues in contact with the stent [8, 12]. Compared to other alloys, electropolishing of NiTi is more difficult as the electropolishing results strongly depend on the initial surface topology and characteristics and may differ for martensitic and austenitic NiTi [219]. If electropolishing is prolonged, while the surface quality might not significantly be improved, a wavy structure develops [219, 222]. The best polishing results are observed for MRR in the range of 2–4 µm/min depending on the electrolyte, temperature, voltage, and material [219, 222].

NiTi is generally considered to be very resistant to most acids and seawater because of the stable oxide layer [8], and therefore, a surface modification to prevent NiTi from corrosion is not required in many cases. For biomedical applications, however, it is mandatory to treat the surface. Due to the high nickel content, the use of NiTi generally exhibits a risk in biomedical applications because nickel can lead to inflammation or tissue reaction. However, it is a general belief that preferential formation of titanium oxide on the surface enhances biocompatibility, passivity, and corrosion resistance of NiTi. This is because the stable surface layer, which consists of more than 95% of oxides, only has a low nickel concentration, and therefore, the risk of nickel contact with the surrounding tissue as well as the release of nickel is reduced. Essentially, the oxide layer acts as a barrier.

Principally, electropolished surfaces however provide a better biocompatibility than mechanically polished or untreated surfaces. The constitution and thickness of the oxide layer strongly affect the biocompatibility in contact with body fluids [8]. In electropolishing, the initial oxide layer is removed, but a new layer is formed immediately because of the high titanium content. This newly formed layer is more homogeneous and uniform than the former layer and prevents the underlying material more effectively from corrosion and from nickel release [219, 221]. Therefore, electropolishing usually is applied to biomedical NiTi devices and further improvement of biocompatibility can be achieved by passivation following the electropolishing step [12]. It is beyond the scope of this chapter to describe the influence of surface treatments for biomedical NiTi applications. There are a tremendous number of researches in this area with some reported contradictory results. The comprehensive work of Shabalovskaya et al. is recommended for further reading [220, 223, 224].

In addition to pure cleaning, smoothing, and enhancement of biocompatibility, surface finishing technologies provide potential for improving physical and mechanical properties, especially the fatigue life [12]. Since structural fatigue is usually initiated from the surface of a part, a smooth surface is beneficial for parts subjected to cyclic or periodic loads. Fatigue performances of NiTi wires improves by nearly 300% after electropolishing or chemical etching compared to untreated NiTi [225]. However, highest improvement is achieved by mechanical polishing because mechanical polishing creates residual compressive stresses in surface regions which provide a crack resistant layer and stop crack propagation.

Coating technologies are applied to NiTi parts for enhanced corrosion properties or for esthetic purposes. For example, orthodontic wires are coated with gold or polymers [8]. Metals are usually deposited by sputtering, by spray coating, or by electrolytic or electroplating processes [12]. Nevertheless, there are problems reported when coating NiTi. During plating, hydrogen charged into the alloy can lead to embrittlement of NiTi [12, 33]. Also, a high ductility of the coating material is required to prevent flaking or cracking when the

NiTi part operates in shape memory or pseudoelastic regimes because otherwise, the strain at the plated surface may be beyond what the plating can tolerate [12, 33]. The adhesion and integrity of the coating is mainly affected by the surface topology and characteristics of the underlying NiTi. The surface oxide layer on NiTi is extremely tenacious which makes it challenging to achieve a good metal-to-metal bond and often leads to poor adhesion at the interface [33]. Crack initiation and crack growth can occur in coated NiTi subjected to comparatively low strains [226]. Highly biocompatible, industrially available diamond-like carbon and titanium coatings as well as nitrided surfaces are not able to withstand strain levels which developed during martensitic transformation in superelastic NiTi. In contrast to uncoated NiTi, cracks in nitrided surfaces are detected at only 1% local strain and at 3% local strain in diamond-like carbon coated NiTi.

References

[1] Saburi T 1998 Ti–Ni shape memory alloys. In Otsuka K and Wayman C M (eds.): *Shape Memory Materials*, Cambridge, UK: Cambridge University Press, pp. 49–96.

[2] Zhang Z, Frenzel J, Somsen C, Pesicka J, and Eeggeler G 2005 On the formation of TiC crystals during processing of NiTi shape memory alloys. In Karas G (ed.): *Trends in Crystal Growth Research*, New York: Nova Science Publishers, pp. 71–99.

[3] Zhang Z, Frenzel J, Somsen C, Pesicka J, Neuking K, and Eggeler G 2006 Orientation relationship between TiC carbides and B2 phase in as-cast and heat-treated NiTi shape memory alloys *Mater. Sci. Eng. A* **438–440** 879–82.

[4] Frenzel J, Zhang Z, Somsen C, Neuking, K, and Eggeler G 2007 Influence of carbon on martensitic phase transformations in NiTi shape memory alloys *Acta Mater.* **55**(4) 1331–41.

[5] Frenzel J, Neuking K, Eggeler G, and Haberland C 2008 *On the role of carbon during processing of NiTi shape memory alloys*. In Miyazaki S (ed.): Proceedings of the International Conference on Shape Memory and Superelastic Technologies, Tsukuba, Japan, December 2–5, 2007, Materials Park, OH: ASM International, pp. 131–8.

[6] Nevitt M V 1960 Stabilization of certain Ti_2Ni-type phases by oxygen *Trans. Metall. Soc. AIME* **218** 327–31.

[7] Olier P, Barcelo F, Bechade J L, Brachet J C, Lefevre E, and Guenin G 1997 Effects of impurities content (oxygen, carbon, nitrogen) on microstructure and phase transformation temperatures of near equiatomic TiNi shape memory alloys *J. Phys. IV* **7**(C5) 143–8.

[8] Mertmann M 2004 Herstellungs- und Verarbeitungseigenschaften von NiTi-Formgedächtnislegierungen. In Gümpel P, Gläser S, Jost N, Mertmann M, Seitz N, and Strittmatter J (eds.): *Formgedächtnislegierungen—Einsatzmöglichleiten in Maschinenbau, Medizintechnik und Aktuatorik*, Renningen: Expert Verlag, pp. 34–54.

[9] Russel S M 2001 *Nitinol melting and fabrication*. In Russel S M and Pelton A R (eds.): Proceedings of the International Conference on Shape Memory and Superelastic Technologies, Pacific Grove, CA, USA, April 30–May 4, 2000, Carrboro, NC: TIPS Technical Publishing Inc., pp. 1–9.

[10] Frenzel J, Zhang Z, Neuking K, and Eggeler G 2004 High quality vacuum induction melting of small quantities of NiTi shape memory alloys in graphite crucibles *J. Alloys Compd.* **385** 214–23.

[11] Frenzel J, Neuking K, and Eggeler G 2004 Induction melting of NiTi shape memory alloys—The influence of the commercial crucible graphite on alloy quality *Mater. Werkst.* **35**(5) 352–8.

[12] Wu M H 2002 *Fabrication of nitinol materials and components*. In Chu Y Y and Zhao L C (eds.): Proceedings of the International Conference on Shape Memory and Superelastic Technologies, Kunming, China, September 3–6, 2001, Mater. Sci. Forum 394–395, pp. 285–92.

[13] Suzuki Y 1998 Fabrication of shape memory alloys. In Otsuka K and Wayman C M (eds.): *Shape Memory Materials*, Cambridge, UK: Cambridge University Press, pp. 133–48.

[14] Wang L M, Liu L H, Yang H, Wang L Y, and Xiu G Q 2002 *Melting and fabrication of NiTi shape-memory alloy wires*. In Chu Y Y and Zhao L C (eds.): Proceedings of the International Conference on Shape Memory and Superelastic Technologies, Kunming, China, September 3–6, 2001, Mater. Sci. Forum 394–395, pp. 297–300.

[15] Drennen D C, Jackson C M, and Wagner H J 1968 The development of melting and casting procedures for nitinol nickel-base alloys. Sandia Laboratories, Albuquerque, SC-CR-69-3070. Battelle Memorial Institute, Columbus, OH.

[16] Grossmann C, Frenzel J, Sampath V, Depka T, Oppenkowski A, Somsen C, Neuking K, Theisen W, and Eggeler G 2008 Processing and property assessment of NiTi and NiTiCu shape memory actuator springs *Mater. Werkst.* **39**(8) 499–510.

[17] Frenzel J, George E P, Dlouhy A, Somsen C, Wagner M F X, and Eggeler G 2010 Influence of Ni on martensitic phase transformation in NiTi shape memory alloys *Acta Mater.* **58** 3444–58.

[18] Coda A, Zillo S, Norwich D, and Sczerzenie F 2012 Characterization of inclusions in VIM/VAR NiTi alloys *J. Mater. Eng. Perform.* **21**(12) 2572–7.

[19] Otubo J, Rigo O D, Neto C d M, Kaufman M J, and Mei P R 2004 Low carbon content NiTi shape memory alloy produced by electron beam melting *Mater. Res.* **7**(2) 263–7.

[20] Young M L, DeFouw J D, Frenzel J, and Dunand D C 2012 Cast-replicated NiTiCu foams with superelastic properties *Metall. Mater. Trans. A* **43** 2939–44.

[21] Sugiyama M, Hyun S K, Tane M, and Nakajima H 2011 Fabrication of lotus-type porous NiTi shape memory alloys using the continuous zone melting method and tensile property *High Temp. Mater. Processes* **26**(4) 297–302.

[22] Duerig T W 1995 NiTi alloys produced by powder metallurgical methods. In Pelton A R, Hodgson D, and Duerig T W (eds.): Proceedings of the 1st International Conference on Shape Memory and Superelastic Technologies, Pacific

Grove, CA, USA, March 7–10, 1994, Monterey, CA: Monterey Institute for Advanced Studies, pp. 31–42.

[23] Prokoshkin S D, Brailovski V, Inaekyan K E, Demers V, Khmekevskaya I Y, Dobatkin S V, and Tatyanin E V 2008 Structure and properties of severely cold-rolled and annealed Ti–Ni shape memory alloys *Mater. Sci. Eng. A* **481–482** 114–8.

[24] Burow J, Prokofiev E, Somsen C, Frenzel J, Valiev R Z, and Eggeler G 2008 Martensitic transformations and functional stability in ultra-fine grained NiTi shape memory alloys *Mater. Sci. Forum* **584–586** 852–7.

[25] Maaß B, Burow J, Frenzel J, and Eggeler G 2009 On the influence of crystal defects on the functional stability of NiTi based shape memory alloys. In Šittner P, Paidar V, Heller L, and Seiner H (eds.): Proceedings of the 8th European Symposium on Martensitic Transformations, ESOMAT 2009, Prague, Czech Republic, September 7–11, 2009, Prague: EDP Sciences, pp. 02022-1–9.

[26] Frenzel J, Burow J A, Payton E J, Rezanka S, and Eggeler G 2011 Improvement of NiTi shape memory actuator performance through ultra-fine grained and nanocrystalline microstructures *Adv. Eng. Mater.* **13**(4) 256–68.

[27] Miller D A and Lagoudas D C 2001 Influence of cold work and heat treatment on the shape memory effect and plastic strain development of NiTi *Mater. Sci. Eng. A* **308** 161–75.

[28] Mitwally M E and Farag M 2009 Effect of cold work and annealing on the structure and characteristics of NiTi alloy *Mater. Sci. Eng. A* **519** 155–66.

[29] Aslanidis D E and Van Moorleghem W 2004 *Process optimization towards high grade wire and tubing surfaces.* In Pelton A R and Duerig T (eds.): Proceedings of the International Conference on Shape Memory and Superelastic Technologies, Pacific Grove, CA, USA, May 5–8, 2003, Menlo Park, CA: SMST Society, Inc., pp. 119–27.

[30] Tuissi A, Bassani P, Mangioni A, Toia L, and Butera F 2006 *Fabrication process and characterization of NiTi wires for actuators.* In Mertmann M (ed.): Proceedings of the International Conference on Shape Memory and Superelastic Technologies, Baden-Baden, Germany, October 3–7, 2004, Materials Park, OH: ASM International, pp. 501–8.

[31] Pelton A R, Dicello J, and Miyazaki S 2000 Optimization of processing and properties of medical grade nitinol wire *Minim. Invasive Ther.* **9**(2) 107–18.

[32] Wu S K, Lin H C, and Chen C C 1999 A study on the machinability of a $Ti_{49.6}Ni_{50.4}$ shape memory alloy *Mater. Lett.* **40** 27–30.

[33] Hodgson D E 2001 Fabrication, heat treatment and joining of nitinol components. In Russel S M and Pelton A R (eds.): Proceedings of the International Conference on Shape Memory and Superelastic Technologies, Pacific Grove, CA, USA, April 30–May 4, 2000, Carrboro, NC: TIPS Technical Publishing Inc., pp. 11–24.

[34] Weinert K and Petzoldt V 2006 Micromachining of NiTi shape memory alloys *Prod. Eng. – Res. Develop.* **13**(2) 43–6.

[35] Biermann D, Kahleyß F, Krebs E, and Upmeier T 2011 A study on micromachining technology for the machining of NiTi: Five-axis micro-milling and deep-hole drilling *J. Mater. Eng. Perform.* **20**(4–5) 745–51.

[36] Frotscher M, Kahleyß F, Simon T, Biermann D and Eggeler G 2011 Achieving small structures in thin NiTi sheets for medical applications with water jet and micro machining: A comparison *J. Mater. Eng. Perform.* **20**(4–5) 776–82.

[37] Theisen W and Schuermann A 2004 Electro discharge machining of nickel–titanium shape memory alloys *Mater. Sci. Eng. A* **378** 200–4.

[38] Stöckel D 2001 Forming nitinol—A challenge. In Siegert K (ed.): *New Developments in Forging Technology,* Frankfurt/Main: MAT-INFO-Werkstoff-Informationsgesellschaft mbH, pp. 119–34.

[39] Schetky L McD and Wu M H 2004 *Issues in the further development of nitinol properties and processing for medical device applications.* In Shrivastava S (ed.): Medical Device Materials—Proceedings of the Materials and Processes for Medical Devices Conference, Anaheim, CA, USA September 8–10, 2003, Materials Park, OH: ASM International, pp. 271–6.

[40] Manjaiah M, Narendranath S, and Basavarajappa S 2014 Review on non-conventional machining of shape memory alloys *Trans. Nonferrous Met. Soc. China* **24** 12–21.

[41] Chen S L, Hsieh S F, Lin H C, Lin M H, and Huang J S 2007 Electrical discharge machining of TiNiCr and TiNiZr ternary shape memory alloys *Mater. Sci. Eng. A* **445–446** 486–92.

[42] Huang H, Zheng H Y, and Liu Y 2005 Experimental investigations of the machinability of $Ni_{50.6}Ti_{49.4}$ alloy *Smart Mater. Struct.* **14** 297–301.

[43] Schüssler A and Strobel M 2003 *Status and trends of nitinol micromachining techniques.* In: Pelton A R and Duerig T (eds.): Proceedings of the International Conference on Shape Memory and Superelastic Technologies, Pacific Grove, CA, USA, May 5–8, 2003, Menlo Park, CA: SMST Society, Inc., pp. 135–41.

[44] Frotscher M, Schreiber F, Neelakantan L, Gries T, and Eggeler G 2011 Processing and characterization of braided NiTi microstents for medical applications *Mater. Werkst.* **42**(11) 1002–12.

[45] Kauffman G B and Mayo I 1996 The story of Nitinol: The serendipitous discovery of the memory metal and its applications *Chem. Educ.* **2**(2) 1–12.

[46] Stöckel D, Pelton, A, and Duerig T 2004 Self-expanding nitinol stents: Material and design considerations *Eur. Radiol.* **14** 292–301.

[47] Heller L, Vokoun D, Majtás D, and Šittner P 2008 Thermomechanical characterization of shape memory alloy tubular composite structures *Adv. Sci. Technol.* **59** 150–5.

[48] Ahlhelm F, Kaufman R, Ahlhelm D, Ong M F, Roth C, and Reith W 2009 Carotid artery stenting using novel self-expanding braided nickel–titanium stent: Feasibility and safety porcine trail *Cardiovasc. Intervent. Radiol.* **32** 1019–27.

[49] Breidert J, Michutta J, Siegert W, Neuking K, and Welp E G 2004 *Soldered joints for shape memory components.* In Pelton A R and Duerig T (eds.): Proceedings of the International Conference on Shape Memory and Superelastic Technologies, Pacific Grove, CA, USA, May 5–8, 2003, Menlo Park, CA: SMST Society Inc. pp. 209–17.

[50] Qiu X M, Li M G, Sun D Q, and Liu W H 2006 Study on brazing of TiNi shape memory alloy with stainless steels *J. Mater. Process. Technol.* **176**(1–3) 8–12.

[51] Li M G, Sun D Q, Qiu X M, Sun D X, and Yin S Q 2006 Effects of laser brazing parameters on microstructure and properties of TiNi shape memory alloy and stainless steel joint *Mater. Sci. Eng. A* **424**(1–2) 17–22.

[52] Shaw J A and Grummon D S 2011 Manufacture of shape memory alloy cellular materials and structures by transient-liquid reactive joining. US Patent No. US 7,896,222 B2.

[53] Hall P C 2000 Resistance welding Ti-rich nitinol wire. In Russel S M and Pelton A R (eds.): Proceedings of the International Conference on Shape Memory and Superelastic Technologies, Pacific Grove, CA, USA, April 30–May 4, 2000, Carrboro, NC: TIPS Technical Publishing Inc., pp. 67–75.

[54] Tam B 2010 *Micro-welding of nitinol shape memory alloy.* University of Waterloo, Waterloo, Canada.

[55] Akselen O M 2010 Joining of shape memory alloys. In: Cismasiu C (ed.): *Shape Memory Alloys*, Rijeka: In Tech, pp. 183–210.

[56] Gugel H 2011 *Laserschweißen artgleicher und artfremder Materialkombinationen mit Nickel-Titan Formgedächtnislegierungen.* Ruhr-Universität Bochum, Bochum, Germany.

[57] Tuissi A, Besseghini S, Ranucci T, Squatrito F, and Pozzi M 1999 Effect of Nd-YAG laser welding on the functional properties of Ni-49.6 at.-%Ti *Mater. Sci. Eng. A* **273–275** 813–7.

[58] Ogata Y, Takatuga M, Kunimasa T, Uenishi K, and Kobayashi K F 2004 Tensile strength and pseudo-elasticity of YAG laser spot melted Ti–Ni shape memory alloy wires *Mater. Trans.* **45**(4) 1070–6.

[59] Tuissi A, Bassani P, Gerosa M, Mauri D, Pini M, Capello E, Previtali B, and Vedani M 2004 CO_2 *laser welding of NiTi/Ni-based alloys.* In Pelton A R and Duerig T (eds.): Proceedings of the International Conference on Shape Memory and Superelastic Technologies, Pacific Grove, CA, USA, May 5–8, 2003, Menlo Park, CA: SMST Society Inc. pp. 229–38.

[60] Falvo A, Furgiuele F M, and Maletta C 2005 Laser welding of a NiTi alloy: Mechanical and shape memory behavior *Mater. Sci. Eng. A* **412** 235–40.

[61] Song Y G, Li W S, Li L, and Zheng Y F 2008 The influence of laser welding parameters on the microstructure and mechanical property of the as-jointed NiTi alloy wires *Mater. Lett.* **62** 2325–8.

[62] Gugel H Schuermann A, and Theisen W 2008 Laser welding of NiTi wires *Mater. Sci. Eng. A* **481–482** 668–71.

[63] Falvo A, Furgiuele F M, and Maletta C 2008 Functional behavior of a NiTi-welded joint: To-way shape memory effect *Mater. Sci. Eng. A* **481–482** 647–50.

[64] Chan C W, Man H C, and Yue T M 2011 Effects of process parameters upon the shape memory and pseudo-elastic behaviors of laser welded NiTi thin foils *Metall. Mater. Trans. A* **42** 2264–70.

[65] Pouquet J, Miranda R M, Quintino L, and Williams S 2012 Dissimilar laser welding of NiTi to stainless steel *Inter. J. Adv. Manuf. Technol.* **61**(1–4) 205–12.

[66] Chan C W, Man H C, and Yue T M 2012 Effect of postweld heat treatment on the microstructure and cyclic deformation behavior of laser-welded NiTi shape memory wires *Metall. Mater. Trans. A* **43** 1956–65.

[67] Hall P C 2004 *Laser welding nitinol to stainless steel*. In Pelton A R and Duerig T (eds.): Proceedings of the International Conference on Shape Memory and Superelastic Technologies, Pacific Grove, CA, USA, May 5–8, 2003, Menlo Park, CA: SMST Society Inc., pp. 219–28.

[68] Li H M, Sun D Q, Cai X L, Dong P, and Wang W Q 2012 Laser welding of TiNi shape memory alloy and stainless steel using Ni interlayer *Mater. Des.* **39** 285–93.

[69] Li H, Sun D, Cai X, Dong P, and Gu X 2013 Laser welding of TiNi shape memory alloy and stainless steel using Co filler metal *Opt. Laser Technol.* **45** 453–60.

[70] Bram M, Ahmad-Khanlou A, Heckmann A, Fuchs B, Buchkremer H P, and Stöver D 2002 Powder metallurgical fabrication processes for NiTi shape memory alloy parts *Mater. Sci. Eng. A* **337** 254–63.

[71] Butler J, Tiernan P, Gandhi A A, Mcnamara K, and Tofail S A M 2011 Production of NiTinol wire from elemental nickel and titanium powders through spark plasma sintering and extrusion *J. Mater. Eng. Perform.* **20**(4–5) 757–61.

[72] German R M 1989 *Particle Packing Characteristics*, Princeton, NJ: Metal Powder Industries Federation.

[73] German R M 2005 *Powder Metallurgy & Particulate Materials Processing*, Princeton, NJ: Metal Powder Industries Federation.

[74] Haberland C 2012 *Additive Verarbeitung von NiTi-Formgedächtniswerstoffen mittels selective laser melting*. Ruhr University Bochum, Bochum, Germany. Aachen: Shaker Verlag.

[75] Nanoval 2014 Description of Nanoval process, http://www.nanoval.de/verfahren_eng.htm, accessed on February 2, 2014.

[76] Köhl M 2009 *Pulvermetallurgie hochporöser NiTi-Legierungen für Implantat- und Dämpfungsanwendungen*. Ruhr University Bochum, Bochum, Germany. Jülich: Forschungszentrum Jülich GmbH.

[77] TLS 2014 Description of EIGA process, http://www.tls-technik.de/e_2.html, accessed on February 2, 2014.

[78] Bram M, Bitzer M, Buchkremer H P, and Stöver D 2012 Reproducibility study of NiTi parts made by metal injection moulding *J. Mater. Eng. Perform.* **21**(12) 2701–12.

[79] Li B Y, Rong L J, and Li Y Y 1998 Porous NiTi alloy prepared from elemental powder sintering *J. Mater. Res.* **13**(10) 2847–51.

[80] Li B, Rong L, and Li Y 1999 Microstructure and superelasticity of porous NiTi alloy *Sci. China Ser. E* **42**(1) 94–9.

[81] Li B Y, Rong L J, Luo X H, and Li Y Y 1999 Transformation behavior of sintered porous NiTi alloys *Metall. Mater. Trans. A* **30** 2753–6.

[82] Li B Y, Rong L J, Gjunter V E, and Li Y Y 2000 Porous Ni–Ti shape memory alloys produced by two different methods *Z. Metallkd.* **91**(4) 291–5.

[83] Zhu S L, Yang H J, Fu D H, Zhang L Y, Li C Y, and Cui Z D 2005 Stress-strain behavior of porous NiTi alloys prepared by powder sintering *Mater. Sci. Eng. A* **408** 264–8.

[84] Khalifehzadeh R, Forouzan S, Arami H, and Sadrnezhaad S K 2007 Prediction of the effect of vacuum sintering conditions on porosity and hardness of porous NiTi shape memory alloys using ANFIS *Comput. Mater. Sci.* **40** 359–65.

[85] Schüller E, Bram M, Buchkremer H P, and Stöver D 2004 Phase transformation temperatures for NiTi alloys prepared by powder metallurgical processes *Mater. Sci. Eng. A* **378** 165–9.

[86] Bansiddhi A and Dunand D C 2009 Shape-memory NiTi-Nb foams *J. Mater. Res.* **24**(6) 2107–17.

[87] Zhang Y P, Li D S, and Zhang X P 2007 Gradient porosity and large pore size NiTi shape memory alloys *Scr. Mater.* **57** 1020–3.

[88] Xiong J Y, Li Y C, Wang P D, Hodgson P D, and Wen C E 2008 Titanium–nickel shape memory alloy foams for bone tissue engineering *J. Mech. Behav. Biomed. Mater.* **1** 269–73.

[89] Li D S, Zhang Y P, Ma X, and Zhang X P 2009 Space-holder engineered porous NiTi shape memory alloys with improved pore characteristics and mechanical properties. *J. Alloys Compd.* **474** L1–5.

[90] Bansiddhi A and Dunand D C 2010 Processing of NiTi foams by transient liquid phase sintering *J. Mater. Eng. Perform.* **20**(4–5) 511–6.

[91] Zhao Y, Taya M, Kang Y, and Kawaski A 2005 Compression behaviour of porous NiTi shape memory alloy *Acta Mater.* **53** 337–43.

[92] Majkic G, Chennoufi Y C, Chen Y C, and Salama K 2007 Synthesis of NiTi by low electrothermal loss spark plasma sintering *Metall. Mater. Trans. A* **38** 2523–30.

[93] Neurohr A J, Dunand D C 2011 Shape-memory NiTi with two-dimensional networks of micro-channels *Acta Biomater.* **7** 1862–72.

[94] Barrabés M, Sevilla P, Planell J A, and Gil F J 2008 Mechanical properties of nickel–titanium foams for reconstructive orthopaedics *Mater. Sci. Eng. C* **28** 23–7.

[95] Bansiddhi A and Dunand D C 2007 Shape-memory NiTi foams produced by solid-state replication with NaF *Intermetallics* **15** 1612–22.

[96] Han X, Zou W, Wang R, Jin S, Zhang Z, Li T, and Yang D 1997 Microstructure of TiNi shape-memory alloy synthesized by explosive shock-wave compression of Ti–Ni powder mixture *J. Mater. Sci.* **32** 4723–9.

[97] Chu C L, Chung C Y, Lin P H, and Wang S D 2004 Fabrication of porous NiTi shape memory alloy for hard tissue implants by combustion synthesis *Mater. Sci. Eng. A* **366** 114–9.

[98] Rhalmi S, Charette S, Assad M, Coillard C, and Rivard C H 2007 The spinal cord dura mater reaction to nitinol and titanium alloy particles: A 1-year study in rabbits *Eur. Spine J.* **16** 1063–72.

[99] Gibson L J, Ashby M F, and Harley B A 2010 *Cellular Materials in Nature and Medicine*. Cambridge, UK: Cambridge University Press.

[100] Schüller E M, Hamed O A, Bram M, Buchkremer H P, and Stöver D 2004 *Properties of hot isostatic pressed NiTi components*. In Pelton A R and Duerig T (eds.): Proceedings of the International Conference on Shape Memory and Superelastic Technologies, Pacific Grove, CA, USA, May 5–8, 2003, Menlo Park, CA: SMST Society Inc., pp. 173–82.

[101] Mentz J, Frenzel J, Wagner M F X, Neuking K, Eggeler G, Buchkremer H P, and Stöver D 2008 Powder metallurgical processing of NiTi shape memory alloys with elevated transformation temperatures *Mater. Sci. Eng. A* **491** 270–8.

[102] Bitzer M, Bram M, Buchkremer H P, and Stöver D 2012 Phase transformation behavior of hot isostatically pressed NiTi–X (X = Ag, Nb, W) alloys for functional engineering applications *J. Mater. Eng. Perform.* **21**(12) 2535–45.

[103] Hellotrade International 2014 Actipore device, http://www.hellotrade.com/biorthex-canada/actipore.html, accessed on March 13, 2014.

[104] Likibi F, Assad M, Jarzem P, Leroux M A, Coillard C, Chabot G, and Rivard C H 2004 Osseointegration study of porous nitinol versus titanium orthopaedic implants *Eur. J. Orthop. Surg. Traumatol.* **14** 209–13.

[105] Assad M, Jarzem P, Leroux M A, Coillard C, Chernyshov A V, Charette S, and Rivard C H 2003 Porous titanium–nickel for intervertebral fusion in a sheep model: part 1. Histomorphometric and radiological analysis *J. Biomed. Mater. Res. B* **64**(2) 107–20.

[106] Lagoudas D C and Vandygriff E L 2002 Processing and characterization of NiTi porous SMA by elevated pressure sintering *J. Intell. Mater. Syst. Struct.* **13** 837–50.

[107] Wu S, Chung C Y, Liu X, Chu P K, Ho J P Y, Chu C L, Chan Y L, Yeung K W K, Lu W W, Cheung K M C, and Luk K D K 2007 Pore formation mechanism and characterization of porous NiTi shape memory alloys synthesized by capsule-free hot isostatic pressing *Acta Mater.* **55** 3437–51.

[108] Wu S L, Liu X M, Chua P K, Chung C Y, Chu C L, and Yeung K W K 2008 Phase transformation behavior of porous NiTi alloys fabricated by capsule-free hot isostatic pressing. *J. Alloys Compd.* **449** 139–43.

[109] Krone L 2005 *Metal-injection-moulding (MIM) von NiTi Bauteilen mit Formgedächtniseigenschaften.* Ruhr University Bochum, Bochum, Germany.

[110] Schatt W, Wieters K P, and Kieback B 2007 *Pulvermetallurgie—Technologie und Werkstoffe*, 2nd edition. Berlin/Heidelberg: Springer-Verlag.

[111] Greiner C, Oppenheimer S M, and Dunand D C 2005 High strength, low stiffness, porous NiTi with superelastic properties *Acta Biomater.* **1** 705–16.

[112] Yuan B, Chung C Y, Huang P, and Zhu M 2006 Superelastic properties of porous TiNi shape memory alloys prepared by hot isostatic pressing *Mater. Sci. Eng. A* **438–440** 657–60.

[113] Yuan B, Chung C Y, Huang P, and Zhu M 2006 The effect of porosity on phase transformation behavior of porous Ti–50.8 at.-% Ni shape memory alloys prepared by capsule-free hot isostatic pressing *Mater. Sci. Eng. A* **438–440** 585–8.

[114] Yuan B, Chung C, and Zhu M 2004 Microstructure and martensitic transformation behavior of porous NiTi shape memory alloy prepared by hot isostatic pressing processing *Mater. Sci. Eng. A* **382** 181–7.

[115] Zhang Y P, Yuan B, Zeng M Q, Chung C Y, and Zhang X P 2007 High porosity and large pore size shape memory alloys fabricated by using pore-forming agent (NH4HCO3) and capsule-free hot isostatic pressing *J. Mater. Process. Technol.* **192–193** 439–42.

[116] Bansiddhi A and Dunand D C 2008 Shape-memory NiTi foams produced by replication of NaCl space-holders *Acta Biomater.* **4** 1996–2007.

[117] Köhl M, Bram M, Moser A, Buchkremer H P, Beck T, and Stöver D 2011 Characterization of porous, net-shaped NiTi alloy regarding its damping and energy-absorbing capacity *Mater. Sci. Eng. A* **528** 2454–62.

[118] Krone L, Mentz J, Bram M, Buchkremer H P, Stöver D, Wagner M, Eggeler G, Christ D, Reese S, Bogdanski D, Köller M, Esenwein S A, Muhr G, Prymak O, and Epple M 2005 The potential of powder metallurgy for the fabrication of biomaterials on the basis of nickel–titanium: A case study with a staple showing shape memory behaviour *Adv. Eng. Mater.* **7** 613–9.

[119] Mentz J, Bram M, Buchkremer H P, and Stöver D 2008 Influence of heat treatment on the mechanical properties of high-quality Ni-rich NiTi produced by powder metallurgical methods *Mater. Sci. Eng. A* **481–482** 630–4.

[120] Bram M, Köhl M, Buchkremer H P, and Stöver D 2011 Mechanical properties of highly porous NiTi alloys *J. Mater. Eng. Perform.* **20**(4–5) 522–8.

[121] Köhl M, Habijan T, Bram M, Buchkremer H P, Stöver D, and Köller M 2009 Powder metallurgical near-net-shape fabrication of porous NiTi shape memory alloys for use as long-term implants by the combination of the metal injection molding process with the space-holder technique *Adv. Eng. Mater.* **11**(12) 959–68.

[122] Köhl M, Bram M, Buchkremer H P, Stöver D, Habijan T, and Köller M 2008 Powder metallurgical production, mechanical and biomedical properties of porous NiTi shape memory alloys. In Gilbert J (ed.): Medical Device Materials IV: Proceedings from the Materials & Processes for Medical Devices Conference, Palm Desert, CA, USA, September 23–27, 2007. Material Park, OH: ASM International, pp. 14–9.

[123] Schüller E M, Bram M, Buchkremer H P, and Stöver D 2004 *Metal injection molding for NiTi alloys*. In Pelton A R and Duerig T (eds.): Proceedings of the International Conference on Shape Memory and Superelastic Technologies, Pacific Grove, CA, USA, May 5–8, 2003, Menlo Park, CA: SMST Society Inc., pp. 143–52.

[124] Hu G, Zhang L, Fan Y, and Li Y 2007 Fabrication of high porous NiTi shape memory alloy by metal injection molding *J. Mater. Process. Technol.* **206** 395–9.

[125] Shishkovsky I V 2005 Shape memory effect in porous volume NiTi articles fabricated by selective laser sintering *Tech. Phys. Lett.* **31** 15–21.

[126] Malukhin K and Ehmann K 2006 Material characterization of NiTi based memory alloys fabricated by laser direct metal deposition process *J. Manuf. Sci. Eng.* **128** 691–6.

[127] Chalker P R, Clare A T, Davies S, Sutcliffe C J, and Tsopanos S 2006 *Selective laser melting of high aspect ratio 3D nickel–titanium structures for MEMS applications*. In Bull S J (ed.): Surface Engineering for Manufacturing Applications—Materials Research Society Symposium Proceedings, Boston, MA, USA, November 28–December 01, 2005, Warrendale, PA: Materials Research Society, pp. 93–8.

[128] Shishkovsky I V, Morozov Y, and Sumurov I 2007 Nanofractal surface structure under laser sintering of titanium and nitinol for bone tissue engineering *Appl. Surf. Sci.* **254**(4) 1145–9.

[129] Yang Y, Huang Y, and Wu W 2007 One-step shaping of NiTi biomaterial by selective laser melting. In Deng S (ed.): ELasers in Materials Processing and Manufacturing 3—Proceedings of the SPIE 6825, Bellingham, WA: Society of Photo-Optical Instrumentation Engineers, pp. 68250C-1-7.

[130] Krishna B V, Bose S, and Bandyopadhyay A 2007 Laser processing of net-shape NiTi shape memory alloy *Metall. Mater. Trans. A* **38** 1096–103.

[131] Shishkovsky I V, Volova L T, Kuznetsov M V, Morozov, Y G, and Parkin I P 2008 Porous biocompatible implants and tissue scaffolds synthesized by selective laser sintering from Ti and NiTi *J. Mater. Chem.* **18** 1309–17.

[132] Clare A T, Chalker P R, Davies S, Sutcliffe C J, and Tsopanos S 2008 Selective laser melting of high aspect ratio 3D nickel-titanium structures two way trained for MEMS applications *Int. J. Mech. Mater. Des.* **4** 181–7.

[133] Meier H, Haberland C, Frenzel J, and Zarnetta R 2009 Selective laser melting of NiTi shape memory components. In Bártolo P J et al. (eds.): *Innovative Developments in Design and Manufacturing—Advanced Research in Virtual and Rapid Prototyping*. London: Taylor & Francis, pp. 233–8, http://www.crcpress.com/product/isbn/9780415873079, accessed on April 27, 2015.

[134] Krishna B V, Bose S, and Bandyopadhyay A 2009 Fabrication of porous NiTi shape memory alloy structures using laser engineered net shaping *J. Biomed. Mater. Res. B* **89** 481–90.

[135] Dudziak S, Gieseke M, Haferkamp H, Barcikowski S, and Kracht D 2010 Functionality of laser-sintered shape memory micro-actuators *Phys. Procedia B* **5** 607–15.

[136] Bormann T, Friess S, de Wild M, Schumacher R, Schulz G, and Müller G 2010 Determination of strain fields in porous shape memory alloys using micro computed tomography. In Stock S R (ed.): Developments in X-Ray Tomography 7 – Proceedings of the SPIE 7804, Bellingham, WA: Society of Photo-Optical Instrumentation Engineers, pp. 78041M-1-9.

[137] Habijan T, Haberland C, Meier H, and Köller M 2010 Rapid manufacturing of porous nickel-titanium as a carrier for human mesenchymal stem cells. In Langenbeck's Archives of Surgery 395(6); 14th Annual Meeting on Surgical Research, Rostock, Germany, September 23–25, 2010, Berlin/Heidelberg: Springer-Verlag, p. 772.

[138] Shishokovsky I V, Yadroitsev I A, and Smurov I Y 2011 Selective laser sintering/ melting of NiTinol-hydroxyapatite composite for medical applications *Powder Metall. Met. Ceram.* **50**(5–6) 275–83.

[139] Halani P R and Shin Y C 2011 In situ synthesis and characterization of shape memory nitinol by laser direct deposition *Metall. Mater. Trans. A* **43** 650–7.

[140] Meier H, Haberland C, and Frenzel J 2011 Structural and functional properties of NiTi shape memory alloys produced by selective laser melting. In Bártolo P J

et al. (eds.): *Innovative Developments in Virtual and Physical Prototyping*. London: Taylor & Francis, pp. 291–6, http://www.crcpress.com/product/isbn/ 9780415684187, accessed on April 27, 2015.

[141] Habijan T, Haberland C, Meier H, Frenzel J, Wuwer C, Schildhauer T A, and Köller M 2011 Biocompatibility and particle release of porous nickel-titanium produced by selective laser melting *Biomed. Eng.* **56**(S1).

[142] Dudziak S 2012 Beeinflussung der funktionellen Eigenschaften aktorischer Nickel-Titan-Legierungen durch die aktiven Parameter im Mikrolaserschmelz-prozess. Laser Zentrum Hannover e.V., Germany, Garbsen: PZH Produktionstechnisches Zentrum GmbH.

[143] Haberland C, Meier H, and Frenzel J 2012 On the properties of Ni-rich NiTi shape memory alloys produced by selective laser melting. In Proceedings of SMASIS 2012—Conference on Smart Materials, Adaptive Structures and Intelligent Systems, Stone Mountain, GA, USA, September 19–21, 2012, ASME, pp. 97–104.

[144] Bormann T, Schumacher R, Müller B, Mertmann M, and de Wild M 2012 Tailoring selective laser melting process parameters for NiTi implants *J. Mater. Eng. Perform.* **21**(12) 2519–24.

[145] Shishkovsky I, Yadroitsev I, and Smurov I 2012 Direct selective laser melting of nitinol powder *Phys. Procedia* **39** 447–54.

[146] Haberland C, Elahinia M, Walker J, Meier H, and Frenzel J 2013 Additive manufacturing of complex NiTi shape memory devices and pseudoelastic components. In: Proceedings of SMASIS 2013—Conference on Smart Materials, Adaptive Structures and Intelligent Systems, Snowbird, UT, USA, September 16–19, 2013. New York: ASME.

[147] Haberland C, Walker J, Elahinia M, and Meier H 2013 Visions, concepts and strategies for smart nitinol actuators and complex nitinol structures produced by additive manufacturing. In: Proceedings of SMASIS 2013—Conference on Smart Materials, Adaptive Structures and Intelligent Systems, Snowbird, UT, USA, September 16–19, 2013. New York: ASME.

[148] Walker J, Haberland C, and Elahinia M 2013 An investigation of process parameters on selective laser melting of nitinol. In: Proceedings of SMASIS 2013—Conference on Smart Materials, Adaptive Structures and Intelligent Systems, Snowbird, UT, USA, September 16–19, 2013. New York: ASME.

[149] Habijan T, Haberland C, Meier H, Frenzel J, Wittsiepe J, Wuwer C, Greulich C, Schildhauer T A, and Köller M 2013 The biocompatibility of dense and porous nickel–titanium produced by selective laser melting *Mater. Sci. Eng. C* **33**(1) 419–26.

[150] Halani P R, Kaya I, Shin Y C, and Karaca H E 2013 Phase transformation characteristics and mechanical characterization of nitinol synthesized by laser direct deposition *Mater. Sci. Eng. A* **559** 836–43.

[151] Bormann T, de Wild M, Beckmann F, and Müller B 2013 Assessing the morphology of selective laser melted NiTi-scaffolds for a three-dimensional quantification of the one-way shape memory effect. In Goulbourne N C and Naguib H E

(eds.): Behavior and Mechanics of Multifunctional Materials and Composites—Proceedings of the SPIE 8689, Bellingham, WA: Society of Photo-Optical Instrumentation Engineers, pp. 868914-1-8.

[152] Speirs M, Dadbakhsh S, Buls S, Kruth J P, Van Humbeeck J, Schrooten J, and Luyten J 2013 The effect of SLM parameters on geometrical characteristics of open porous NiTi scaffolds. In Bártolo P J et al. (eds.): *High Value Manufacturing: Advanced Research in Virtual and Rapid Prototyping*. London: Taylor & Francis, pp. 309–14, http://www.crcpress.com/product/isbn/9780415684187, accessed on April 27, 2015.

[153] Wohlers T 2009 *Wohlers Report 2009. State of the Industry—Annual Worldwide Progress Report*. Fort Collins, CO: Wohlers Associates, Inc.

[154] Levi D S, Kusnezov N, and Carman G P 2008 Smart materials applications for pediatric cardiovascular devices *Pediatr. Res.* **63**(5) 552–8.

[155] Otsuka K and Ren X 2005 Physical metallurgy of Ti–Ni-based shape memory alloys *Prog. Mater. Sci.* **50** 511–678.

[156] Pan G, Cao Z, Wei M, Xu L, Shi J, and Meng X 2014 Superelasticity of TiNi thin films induced by cyclic nanoindentation deformation at nanoscale *Mater. Sci. Eng. A* **600** 8–11.

[157] Davies S T, Harvey E C, Jin H, Hayes J P, Ghantasala M K, Roch I, and Buchaillot L 2002 Characterization of micromachining processes during KrF excimer laser ablation of TiNi shape memory alloy thin sheets and films *Smart Mater. Struct.* **11** 708–14.

[158] Fu Y, Du H, Huang W, Zhang S, and Hu M 2004 TiNi-based thin films in MEMS applications: A review *Sensors Actuators A* **112** 395–408.

[159] Sekiguchi Y, Funami K, and Funakubo H 1983 Deposition of NiTi shape memory alloy thin film by vacuum evaporation. In: Proceedings of the 32nd Annual Conference of Japan Society of Materials, pp. 65–7.

[160] Busch J D, Johnson A D, Lee C H, and Stevenson D A 1990 Shape-memory properties in Ni–Ti sputter-deposited film *J. Appl. Phys.* **68** 6224–8.

[161] Gill J J, Chang D T, Momoda L A, and Carman G P 2001 Manufacturing issues of thin film NiTi microwrapper *Sensors Actuators A* **93** 148–56

[162] Kotnur V G, Tichelarr F D, and Jansen G C A M 2013 Sputter deposited Ni–Ti thin films on polyimide substrate *Surf. Coat. Technol.* **222** 44–7.

[163] Matsunaga T, Kajiwara S, Ogawa K, Kikuchi T, and Miyazaki S 1999 High strength Ti–Ni based shape memory thin films *Mater. Sci. Eng. A* **273–275** 745–8.

[164] Ren M H, Wang L, Xu D, and Cai B C 2000 Sputter-deposited Ti–Ni–Cu shaped memory alloy thin films *Mater. Des.* **21** 583–6.

[165] Ramirez A G, Ni H, and Lee H-J 2006 Crystallization of amorphous sputtered thin films *Mater. Sci. Eng. A* **438–440** 703–9.

[166] Mohanchandra K P, Ho K H, and Carman G P 2008 Compositional uniformity in sputter-deposited NiTi shape memory alloy thin films *Mater. Lett.* **62** 3481–3.

[167] Rao J, Roberts T, Lawson K, and Nicholls J 2010 Nickel titanium and nickel titanium hafnium shape memory alloy thin films *Surf. Coat. Technol.* **204** 2331–6.

[168] Miyazaki S and Ishida A 1999 Martensitic transformation and shape memory behavior in sputter-deposited TiNi-base thin films *Mater. Sci. Eng. A* **273–275** 106–33.

[169] Ishida A, Sato M, Kimura T, and Sawaguchi T 2001 Effects of composition and annealing on shape memory behavior of Ti-rich Ti–Ni films formed by sputtering *Mater. Trans.* **42**(6) 1060–7.

[170] Rumpf H, Winzek B, Zamponi C, Siegert W, Neuking K, and Quandt E 2004 Sputter deposition of NiTi to investigate the Ti loss rate as a function of composition from cast melted targets *Mater. Sci. Eng. A* **378** 429–33.

[171] Zamponi C, Rumpf H, Wehner B, Frenzel J, and Quandt E 2004 Superelasticity of free-standing NiTi films depending on the oxygen impurity of the used targets *Mater. Werkst.* **35**(5) 359–64.

[172] Frenzel J, Frotscher M, Petzold V, Neuking K, Eggeler G, and Weinert K 2006 *NiTi shape memory alloy metallurgy—Fabrication of ingots and sputter targets.* In Mertmann M (ed.): Proceedings of the International Conference on Shape Memory and Superelastic Technologies, Baden-Baden, Germany, October 3–7, 2004, Materials Park, OH: ASM International, pp. 223–8.

[173] Ho K K, Gregory P, and Carman G P 2000 Sputter deposition of NiTi thin film shape memory alloy using a heated target *Thin Solid Films* **370** 18–29.

[174] Ho K K, Mohanchandra G P, and Carmann G P 2002 Examination of the sputtering profile of NiTi under target heating conditions *Thin Solid Films* **413** 1–7.

[175] Gupta V, Johnson A D, Martynov V, and Menchaca L 2004 Nitinol thin film three-dimensional devices—Fabrication and applications. In Pelton A R and Duerig T (eds.): Proceedings of the International Conference on Shape Memory and Superelastic Technologies, Pacific Grove, CA, USA, May 5–8, 2003, Menlo Park, CA: SMST Society Inc. pp. 639–50.

[176] Guilemany J M, Cinca N, Dosta S, and Benedetti A V 2009 Corrosion behavior of thermal sprayed nitinol coatings *Corros. Sci.* **51** 171–80.

[177] Cinca N, Isalgué A, Fernándes J, and Guilemany J M 2009 NiTi thermal sprayed coatings characterization. In Šittner P, Paidar V, Heller L, and Seiner H (eds.): Proceedings of the 8th European Symposium on Martensitic Transformations, ESOMAT 2009, Prague, Czech Republic, September 7–11, 2009, Prague: EDP Sciences, pp. 06006-1–6.

[178] Guilemany J M, Cinca N, Dosta S, and Fernández J 2010 Structural characterization of intermetallic NiTi coatings obtained by thermal spray technologies *Mater. Sci. Forum* **636–637** 1084–90.

[179] Verdian M M, Raeissi K, and Salehi M 2010 Corrosion performance of HVOF and APS thermally sprayed NiTi intermetallic coatings in 3.5% NaCl solution *Corros. Sci.* **52** 1052–9.

[180] Jardine A P, Horan Y, and Herman H 1991 Cavitation-erosion resistance of thick-film thermally sprayed NiTi. In Johnson L A, Pope D P, and Stiegler J O (eds.): MRS Proceedings 213—Symposium Q—High Temperature Ordered Intermetallic Alloys IV, MRS Fall Meeting, Boston, MA, USA, November 26—December 1, 1990. Warrendale, PA: MRS, pp. 815–21.

[181] Hiraga H, Inoue T, Kamado S, Kojima Y, Matsunawa A, and Shimura H 2001 Fabrication of NiTi intermetallic compound coating made by laser plasma hybrid spraying of mechanically alloyed powders *Surf. Coat. Technol.* **139** 93–100.

[182] Horlock A J, Sadeghian Z, McCartney D G, and Shipway P H 2005 High-velocity oxyfuel reactive spraying of mechanically alloyed Ni–Ti–C powders *J. Therm. Spray Technol.* **14**(1) 77–84.

[183] Stella J, Schüller E, Heßing C, Hamed O A, Pohl M, and Stöver D 2006 Cavitation erosion of plasma-sprayed NiTi coatings *Wear* **260** 1020–7.

[184] Cinca N, Isalgué A, Fernández J, Cano I G, Dosta S, and Guilemany J M 2009 Wear of NiTi coatings obtained by thermal spraying. In Šittner P, Paidar V, Heller L, and Seiner H (eds.): Proceedings of the 8th European Symposium on Martensitic Transformations, ESOMAT 2009, Prague, Czech Republic, September 7–11, 2009, Prague: EDP Sciences, pp. 06007-1-6.

[185] Halter K, Sickinger A, Siegmann S, and Zysset L 2004 Thermal spray forming of NiTi shape memory alloys. In: Pelton A R and Duerig T (eds.): Proceedings of the International Conference on Shape Memory and Superelastic Technologies, Pacific Grove, CA, USA, May 5–8, 2003, Menlo Park, CA: SMST Society, Inc., pp. 163–72.

[186] Otsuka K and Wayman C M (eds.) 1998 *Shape Memory Materials.* Cambridge, UK: Cambridge University Press.

[187] Funakubo H (ed.) 1987 *Precision Machinery and Robotics, Vol. 1: Shape Memory Alloys.* Amsterdam: CRC Press LLC.

[188] Nishida M, Wayman C M, and Honma T 1986 Precipitation processes in near-equiatomic TiNi shape memory alloys *Metall. Trans. A* **17** 1505–15.

[189] Gall K and Maier H J 2002 Cyclic deformation mechanisms in precipitated NiTi shape memory alloys *Acta Mater.* **50** 4643–57.

[190] Ortega A M, Tyber J, Frick C P, Gall K, and Maier H J 2005 Cast NiTi shape-memory alloys *Adv. Eng. Mater.* **7**(6) 492–507.

[191] Khalil-Allafi J, Dloughý A, and Eggeler G 2002 Ni_4Ti_3-precipitation during aging of NiTi shape memory alloys and its influence on martensitic phase transformations *Acta Mater.* **50** 4255–74.

[192] Bojda O, Eggeler G, and Dloughý A 2005 Precipitation of Ni_4Ti_3-variants in a polycrystalline Ni-rich NiTi shape memory alloy *Scr. Mater.* **53** 99–104.

[193] Ghabchi A, Khalil-Allafi J, Liu X W, Söderberg O, Turunen E, and Hannula S P 2009 Effect of aging and solution annealing on transformation and deformation mechanism of super-elastic $Ni_{50.9\%}$–Ti alloy in nano-scale. In Šittner P, Paidar V, Heller L, and Seiner H (eds.): Proceedings of the 8th European Symposium on Martensitic Transformations, ESOMAT 2009, Prague, Czech Republic, September 7–11, 2009, Prague: EDP Sciences, pp. 02009-1-6.

[194] Delville R, Malard B, Pilch J, Šittner P, and Schryvers D 2010 Microstructure changes during non-conventional heat treatment of thin Ni–Ti wires by pulsed electric current studied by transmission electron microscopy *Acta Mater.* **58** 4503–15.

[195] Saburi T, Tatsumi T, and Nenno S 1982 Effects of heat treatment on mechanical behavior of Ti-Ni alloys *J. de Phys.* **43**(12) C4-261–6.

[196] Miyazaki S, Kimura S, Takei F, Miura T, Otsuka K, and Suzuki Y 1983 Shape memory effect and pseudoelasticity in a Ti–Ni single crystal *Scr. Metall.* **17**(9) 1057–62.

[197] Miyazaki S, Imai T, Igo Y, and Otsuka K 1986 Effect of cyclic deformation on the pseudoelasticity characteristics of Ti–Ni alloys *Metall. Trans. A* **17** 115–20.

[198] Huang X and Liu Y 2001 Effect of annealing on the transformation behavior and superelasticity of NiTi shape memory alloy *Scr. Mater.* **45** 153–60.

[199] Michutta J, Somsen C, Yawny A, Dloughý A, and Eggeler G 2006 Elementary martensitic transformation processes in Ni-rich single crystals with Ni_4Ti_3 precipitates *Acta Mater.* **54**(13) 3525–42.

[200] Wagner M, Frenzel J, and Eggeler G 2006 *Evolution of microstructural parameters during cycling and their effect on mechanical and thermal memory.* In Mertmann M (ed.): Proceedings of the International Conference on Shape Memory and Super-elastic Technologies, Baden-Baden, Germany, October 3–7, 2004, Materials Park, OH: ASM International, pp. 103–9.

[201] Jiang F, Yinong L, Yang H, Li L, and Zheng Y 2009 Effect of ageing treatment on the deformation behavior of Ti–50.9 at.-% Ni *Acta Mater.* **57** 4773–81.

[202] Miyazaki S, Kohiyama Y, Otsuka K, and Duerig T 1990 Effects of several factors on the ductility of Ti–Ni alloy *Mater. Sci. Forum* **56–58** 765–70.

[203] Khalil-Allafi J, Ren X, and Eggeler G 2002 The mechanism of multistage marten-sitic transformation in aged Ni-rich NiTi shape memory alloys *Acta Mater.* **50** 793–803.

[204] Khalil-Allafi J, Eggeler G, Dloughý A, Schmahl W W, and Somsen C 2004 On the influence of heterogeneous precipitation on martensitic transformations in a Ni-rich NiTi shape memory alloy *Mater. Sci. Eng. A* **378** 148–51.

[205] Sitepu H, Schmahl WW, Khalil-Allafi J, Eggeler G, Dloughý A, Többens D M, and Tovar M 2002 Neutron diffraction phase analysis during thermal cycling of a Ni-rich NiTi shape memory alloy using the Rietveld method *Scr. Mater.* **46** 543–8.

[206] Zhou N, Shen C, Wagner M F-X, Eggeler G, Mills M J, and Wang Y 2010 Effect of Ni_4Ti_3 precipitation on martensitic transformation in Ti–Ni *Acta Mater.* **58** 6685–94.

[207] Dloughý A, Bojda O, Somsen C, and Eggeler G 2008 Conventional and in-situ transmission electron microscopy investigations into multistage martensitic transformations in Ni-rich NiTi shape memory alloys *Mater. Sci. Eng. A* **481–482** 409–13.

[208] Fan G, Chen W, Yang S, Zhu J, Ren X, and Otsuka K 2004 Origin of abnormal multi-stage martensitic transformation behavior in aged Ni-rich NiTi shape memory alloys *Acta Mater.* **52** 4351–62.

[209] Carroll M C, Somsen C, and Eggeler G 2004 Multiple-step martensitic transfor-mation in Ni-rich NiTi shape memory alloys *Scr. Mater.* **50** 187–92.

[210] Bellouard Y, Lehnert T, Bidaux J E, Sidler T, Clavel R, and Gottardt R 1999 Local annealing of complex mechanical devices: A new approach for developing monolithic micro-devices *Mater. Sci. Eng. A* **273–275** 795–8.

[211] Wang X, Bellouard Y, and Vlassak J J 2005 Laser annealing of amorphous NiTi shape memory alloy thin films to locally induce shape memory properties *Acta Mater.* **53** 4955–61.

[212] Welp E G and Langbein S 2008 Survey of the in situ configuration of cold-rolled, nickel-rich NiTi sheets to create variable component functions *Mater. Sci. Eng. A* **481–482** 602–5.

[213] Langbein S 2009 Development of standardised and integrated shape memory components in "one-module" design. In Šittner P, Paidar V, Heller L, and Seiner H (eds.): Proceedings of the 8th European Symposium on Martensitic Transformations, ESOMAT 2009, Prague, Czech Republic, September 7–11, 2009, Prague: EDP Sciences, pp. 07010-1–9.

[214] Daly M, Pequegnat A, Zhou Y, and Khan M 2013 Fabrication of a novel laser-processed NiTi shape memory microgripper with enhanced thermomechanical functionality *J. Intell. Mater. Syst. Struct.* **24**(8) 984–90.

[215] Facchinello Y, Brailovski V, Inaekyan K, Petit Y, and MacThiong J-M 2013 Manufacturing of monolithic superelastic rods with variable properties for spinal correction: Feasibility study *J. Mech. Behav. Biomed. Mater.* **22** 1–11.

[216] Panton B, Michael A, Pequegnat A, Daly M, Zhou Y, and Khan M I 2013 An innovative laser-processed NiTi self-biasing linear actuator. In: Proceedings of SMASIS 2013—Conference on Smart Materials, Adaptive Structures and Intelligent Systems, Snowbird, UT, USA, September 16–19, 2013. New York: ASME.

[217] Koludrovich M 2014, *Design, analysis, and experimental evaluation of a superelastic NiTi minimally invasive thrombectomy device.* The University of Toledo, Toledo, OH.

[218] Liu X, Wang Y, Yang D, and Qi M 2008 The effect of ageing treatment on shape-setting and superelasticity of a nitinol stent *Mater. Charact.* **59**(4) 402–6.

[219] Pohl M, Heßing C, and Frenzel J 2004 Electrolytic processing of NiTi shape memory alloys *Mater. Sci. Eng. A* **378** 191–9.

[220] Shabalovskaya S, Anderegg J, and Van Humbeeck J 2008 Critical overview of nitinol surfaces and their modifications for medical applications *Acta Biomater.* **4** 447–67.

[221] Decker J F, Trépanier C, Vien L, and Pelton A R 2011 The effect of material removal on the corrosion resistance and biocompatibility of nitinol laser-cut and wire-form products *J. Mater. Eng. Perform.* **20**(4–5) 802–6.

[222] Pohl M 2004 *Elektropolieren und Mikrostrukturieren metallischer Werkstoffe.* In Pohl M (ed): Fortschritte in der Metallographie: Proceedings of the 38. Metallographie-Tagung, September 29–October 01, 2004, Berlin, Germany. Frankfurt/Main: Werkstoff-Informationsgesellschaft mbH, pp. 263–8.

[223] Shabalovskaya S A, Tian H, Anderegg J W, Schryvers D U, Carroll W U, and Van Humbeeck J 2009 The influence of surface oxides on the distribution and release of nickel from nitinol wires *Biomaterials* **30** 468–77.

[224] Shabalovskaya S A, Rondelli G C, Undisz A L, Andereda J W, Burleigh T D, and Rettenmayr M E 2009 The electrochemical characteristics of native nitinol surfaces *Biomaterials* **30** 3662–71.

[225] Polinsky M A, Norwich D W, and Wu M H 2008 *A study of the effects of surface modifications and processing on the fatigue properties of NiTi wire*. In Berg A B, Mitchel M R, and Proft J (eds.): Proceedings of the International Conference on Shape Memory and Superelastic Technologies, Pacific Grove, CA, USA, May 7–11, 2006, Materials Park, OH: ASM International, pp. 1–18.

[226] Heßing C, Frenzel J, Pohl M, and Shabalovskaya S 2008 Effect of martensitic transformation on the performance of coated NiTi surfaces *Mater. Sci. Eng. A* **486** 461–9.

7

Experimental Characterization of Shape Memory Alloys

Ali S. Turabi, Soheil Saedi, Sayed Mohammad Saghaian, Haluk E. Karaca and Mohammad H. Elahinia

7.1 Introduction

Characterization of the microstructure and shape memory behavior of shape memory alloys is an essential part of understanding and modeling the behavior of these alloys. This chapter will focus on the methods and procedures to determine the shape memory properties. Initially, thermodynamics of SMAs and methods to find transformation temperatures will be introduced. Then, shape memory effect and superelasticity tests will be discussed. Effects of aging, crystal structure, orientation, and cycling on the shape memory behavior will be briefly introduced. Important parameters and methods for their measurement such as recoverable and irrecoverable strains, critical stress for transformation, and hysteresis of shape memory

Shape Memory Alloy Actuators: Design, Fabrication, and Experimental Evaluation, First Edition. Mohammad H. Elahinia.
© 2016 John Wiley & Sons, Ltd. Published 2016 by John Wiley & Sons, Ltd.

alloys will be explained. Vickers hardness test procedure, indentation methods, and microstructural characterization by the optical microscopy and X-ray diffraction (XRD) method will also be discussed.

7.2 Characterization of Physical Properties

Unusual behavior of SMAs is due to the reversible shape changes by *martensitic phase transformations*. A martensitic transformation is an example of a displacive (diffusionless shear transformation) transition, in which there is cooperative motion of a relatively large number of atoms, each being displaced by only a small distance, and they move in an organized manner relative to their neighbors. This homogeneous shearing of the parent phase creates a new crystal structure, without any compositional change (no diffusion). A simple thermodynamic analysis of the phase transformations is given here. The Gibbs free energy general form is given by

$$\Delta G = \Delta H - T \Delta S \tag{7.1}$$

where ΔH is the enthalpy change, ΔS is the entropy change, and T is temperature. For simplicity, the Gibbs free energies of martensite and austenite can be assumed to be decreasing linearly with temperature. At the intersection of their Gibbs free energy curves, the transforming phases have the same free energy and are in equilibrium at the equilibrium temperature (T_0). Below T_0, martensite has lower free energy, and therefore, it is favored thermodynamically. Above T_0, austenite is stable. In Figure 7.1, energy curves for austenite and martensite (forward) phase transformation are schematized. G_{ch}^P and G_{ch}^M are the chemical energies of austenite and martensite, respectively.

ΔG_{ch}^{p-m} is the chemical driving force for phase transformation from parent phase (austenite) to martensite and ΔG_{ch}^{m-p} is vice versa. Parent phase transforms to martensite and martensite transforms to parent phase when there is a sufficient driving force in the system. When the G_{ch}^M and G_{ch}^P are equal to each other, no transformation is expected since there is no difference (driving force) between the chemical energies of transforming phases. The general thermodynamical equilibrium equations for the forward transformation can be written of the forms [1]

$$\Delta G_{total}^{p-m} = \Delta G_{ch}^{p-m} + \Delta G_{nc}^{p-m} = \Delta G_{ch}^{p-m} + \Delta G_{el}^{p-m} + \Delta G_{irr}^{p-m} \tag{7.2}$$

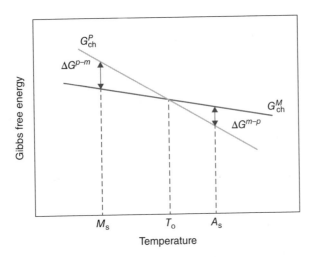

Figure 7.1   Schematics of the free energy curves of martensite and austenite

where ΔG_{total}^{p-m} is the total Gibbs free energy difference to initiate the martensitic transformation, ΔG_{ch}^{p-m} is the change in chemical energy, and ΔG_{nc}^{p-m} is the change in nonchemical energy. ΔG_{nc}^{p-m} energy can be expressed as a combination of ΔG_{el}^{p-m}, the change in elastic energy, and ΔG_{irr}^{p-m}, the irreversible energy during the phase transformation from austenite to martensite [2]. ΔG_{total}^{p-m} term should be smaller than zero in order to initiate the martensitic transformation. At T_0, since there is no driving force to trigger the martensitic transformation, an additional energy should be supplied (by cooling or heating) to initiate the transformation.

Additional cooling $(T_0 - M_s)$ below T_0 is necessary for parent phase to martensite transformation and additional heating $(A_s - T_0)$ beyond T_0 is required for martensite to parent phase transformation assuming negligible elastic energy storage. Shape memory effect and pseudoelasticity/superelasticity originate from the thermoelastic martensitic transformation [3]. ΔG_{el}^{p-m} is the stored elastic energy during the forward transformation, and it is released completely upon back transformation from martensite to austenite. Hence, the elastic energy storage is a reversible process [1]. The amount of the stored elastic energy should be equal to the released energy upon back transformation if there is no plastic relaxation due to dislocation generation/plastic deformation after a full transformation cycle [4, 5].

The irreversible energy ΔG_{irr}^{p-m} can be assumed to stem from two main mechanisms: (i) plastic relaxation energy due to dislocations and defects generation and (ii) friction energy during phase transformation due to the movement of phase transformation front, interaction of martensite variants and internal twins. Both mechanisms result in dissipation energy and thus hysteretic behaviour in SMAs [6].

7.2.1 Differential Scanning Calorimetery

Transformation Temperatures (TTs) are some of the most important shape memory properties. In SMAs, the martensite to austenite transformation (backward transformation) is an endothermic reaction (heat absorbing), while the austenite to martensite transformation (forward transformation) is an exothermic reaction (heat emitting) [7]. The differential scanning calorimeter (DSC) is the most well-known equipment to determine the latent heat, enthalpy, and transformation temperatures [8] (ASTM F2004). The basic principle behind the operation of the DSC is the measurement of the rate at which heat energy is supplied to the specimen in comparison to a reference material to maintain a constant temperature rate [9]. In general, the sample is placed in a pan and then put in one of the holding pans of the furnace, while an empty sample pan is loaded to the other holding for reference. Sample is thermally cycled and the difference of the supplied heat power is recorded as shown in Figure 7.2. The transformation temperatures and enthalpies can be determined from the DSC profiles. Since the transformations occur without

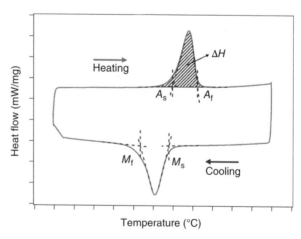

Figure 7.2 Schematic of DSC result and analysis method

any external stress applied, they are also called zero-stress or stress-free transformation temperatures.

Prior to running a DSC test, it is a good idea to calibrate the equipment. The temperature scale is calibrated by using two-point (or more) calibration method where the onset temperatures of the melting points of high-purity material standards (such as zinc and indium) are used. After calibration, a small quantity of the material (10–40 mg) is used for DSC. The SMA sample must be polished to establish good thermal contact with the bottom of the pan, and its weight should be measured. The sample is usually encapsulated in a sample pan and then placed in the holding pan of the furnaces. Both the sample weight and heating rate affect the DSC response since increasing the sample weight and/or heating rate increases the temperature gradients in the sample resulting in decreased signal quality and resolution.

For the most accurate results, the weight of the sample and the heating rate should be kept as low as possible. Thermal cycling rate is generally kept as $10°C/min$, but it can be altered as needed. It should be noted that A_s and M_s stay approximately constant, while A_f and M_f have been observed to decrease with increased temperature rate [10, 11]. The entropy of the martensitic transformation can be calculated by Equation 7.1, and thermodynamic equilibrium can be achieved when ΔG^{m-p} or $\Delta G^{p-m} = 0$ at the equilibrium temperature [12]. Thus, the transformation entropies can be expressed as

$$\Delta S_{m-p} = \frac{\Delta H_{m-p}}{T_0} \qquad (7.3)$$

for the martensite to austenite transformation and

$$\Delta S_{p-m} = \frac{\Delta H_{p-m}}{T_0} \qquad (7.4)$$

for the austenite to martensite transformation. In fact, T_0 is the temperature in which the chemical energies of austenite and martensite phases are balanced [12]. T_0 can be estimated by using the transformation temperature [13, 14]

$$T_0 = \frac{1}{2}(M_s + A_s) \qquad (7.5)$$

or

$$T_0 = \frac{1}{2}(M_f + A_f) \qquad (7.6)$$

The required energy for the sample during the transformation can be obtained through activation energy calculation by using the Kissinger method [15, 16]. The effective activation energy can be determined from the shift of the transformation temperature peaks with heating rate:

$$\frac{d\left(\ln\left(\propto /T_P^2\right)\right)}{d\left(\ln 1/T_P\right)} = \frac{-E}{R} \tag{7.7}$$

where R is the ideal gas constant ($R = 8.314 \, \text{J/mol}$), E is the activation energy, $\propto$ is the rate of heating, and T_P is the temperature for the maximum peak in the DSC curve during transformation. An alternative method to calculate the activation energy has been proposed by Ozawa [17].

Figure 7.2 illustrates a typical DSC response. When the material is heated, transformation begins at A_s (austenite start temperature) and completes at A_f (austenite finish temperature). During cooling, austenite to martensite transformation starts at M_s (martensite start temperature) and ends at M_f (martensite finish temperature). On the calorimetric graph, phase transformations are depicted as peaks, and the areas under those peaks indicate the energies of transformations. Using these peaks, the critical transition temperatures can be determined. The most common method to measure the transformation temperatures is the intersection method. Tangents are drawn at the start and end of the transformation peaks and the base line of the heating and cooling curves. The intersections of those tangent lines are accepted as the transformation temperatures. The enthalpy change of the phase transition can be found by integrating the area between selected temperatures as shown in the Figure 7.2.

In NiTi alloys, cubic austenite (B2) transforms to monoclinic (B19′) martensite. It should be noted that crystal structures and lattice parameters are composition, alloying, and thermomechanical treatment dependent [3, 18–20]. In some cases, multiple phase transformations can be observed. As an example, in NiTi alloys, B2 austenite could transform to R-phase first and then to B19′ martensite. R-phase could appear with increasing Ni content, replacing Ni atoms with a lower 3D element and also, as shown in chapter 6, the result of certain thermomechanical treatments [21].

Figure 7.3 shows the commonly observed lattice structures of NiTi alloys. It is known that adding Cu or Pd to NiTi could result in the formation of orthorhombic (B19) martensite [9]. Figure 7.4 shows the DSC response of an aged Ni-rich NiTi alloy that undergoes B2–R–B19′ transformation during cooling. Upon heating, only B19′–B2 transformation is observed.

It should be noted that multistep martensitic transformations could also be observed in NiTi alloys due to the inhomogeneity of the microstructure.

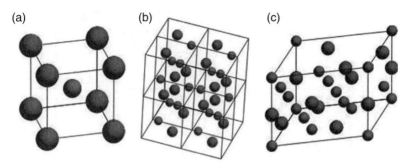

Figure 7.3 The common crystal structures of NiTi: B2 (a), B19 (b), R phase (c). From Ref. [22]

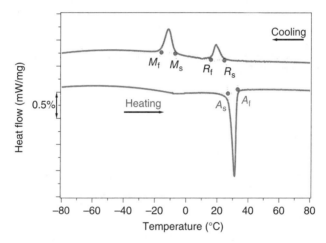

Figure 7.4 R-phase transformation identified on a DSC response of an aged $Ni_{51}Ti_{49}$ alloy

Several factors have been proposed to explain multistage transformation such as stress fields around the precipitates and inhomogeneous distribution of precipitates [23–26]. Figure 7.5 shows the multistep transformation from B2 to R-phase and then to B19' in an aged NiTi alloy. Upon heating, B19' transforms directly back to B2 in some regions (large peak during heating), while B19' transforms to R-phase and then to B2 in other regions (shoulder of the peak).

Figure 7.6 shows the effects of first and second cycles on the transformation temperatures of a cryogenically machined NiTi SMAs. Transformation temperatures in the first cycle are higher than the second cycle since the material was deformed in martensite. The higher transformation temperatures

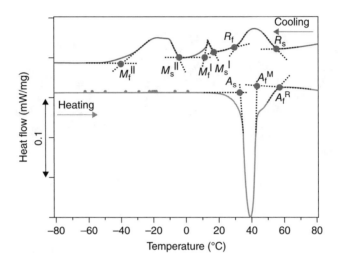

Figure 7.5 Multistep martensitic transformation in aged NiTi alloys. Reproduced with permission from Ref. [27], Elsevier

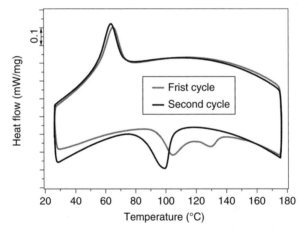

Figure 7.6 First and second DSC cycles of a cryogenically machined NiTi specimen. Reproduced with permission from Ref. [28], Elsevier

in the first cycle can be attributed to formation of the residual stress and high dislocation density and relaxation of elastic energy after machining which suppress and delay the martensite to austenite phase transformation. After the first thermal cycle, peak broadening gets smaller, and hence, both temperatures (A_s and A_f) are reduced and get close to the as-received temperature [28].

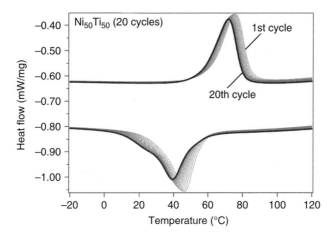

Figure 7.7   Thermal cycling behavior of NiTi

DSC is an efficient tool to conduct thermal cycling experiments. In general, phase transformation temperatures shift to lower temperatures, and peaks become broader with thermal cycling due to plastic deformation, as shown in Figure 7.7. Thermal cycling responses can be related to the strength and the fatigue life of SMAs [29].

7.2.2 Magnetic Measurements

Magnetic measurement is one of the alternative methods to determine the shape memory properties including transformation temperatures and temperature hysteresis. Magnetization of the material can be measured as a function of temperature under an external magnetic field. This method is very effective to reveal the transformation behavior of magnetic shape memory alloys where transforming phases have different magnetization behavior. Transformation temperatures can be measured by tangent method as described for DSC measurements.

Figure 7.8 shows the transformation behavior of $Ni_{45}Mn_{36.5}Co_5In_{13.5}$ single crystalline alloy where the change of its magnetization response with temperature is recorded under constant applied magnetic field [30]. In general, austenite phase is ferromagnetic and martensite phase is weakly magnetic in NiMnCoIn magnetic shape memory alloys. Thus, during cooling, magnetization drops due to forward transformation from highly magnetic austenite to weakly martensite, and back transformation occurs with increasing magnetization level during heating. Temperature cycling under magnetic field

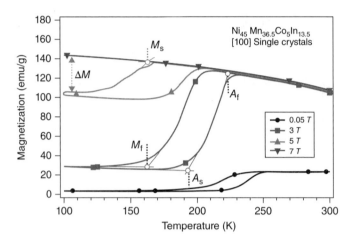

Figure 7.8 Magnetization of NiMnCoIn single crystalline alloys as a function of temperature under selected magnetic fields. Reproduced with permission from Ref. [30], John Wiley & Sons, Ltd

provides the transformation temperatures and hysteresis as a function of applied magnetic field.

7.2.3 Electrical Resistivity

Electrical resistivity measurement is another method to determine the transformation temperatures. In general, a four-point probe method is used to reveal the change of the resistivity of the material with temperature where current from the source passes through between outer probes and voltage is measured between the inner probes. The electrical resistivity of the alloy depends on volume fractions of the existing phases.

A typical electric resistivity measurement of NiTi is shown in Figure 7.9a which can be used to determine the transformation temperatures [31]. During cooling, austenite to R-phase and R-phase to martensite transformation can be detected by the change in resistivity response where resistivity is dramatically increased during austenite to R-phase transformation and decreased during the R-phase to martensite transformation. Back transformation occurs from martensite to austenite directly during heating without substantial change in resistivity. In magnetic shape memory alloys, the resistivity is both temperature and magnetic field dependent where Figure 7.9b shows the electrical resistivity change during thermal cycling under constant magnetic field in NiMnCoIn alloy [32].

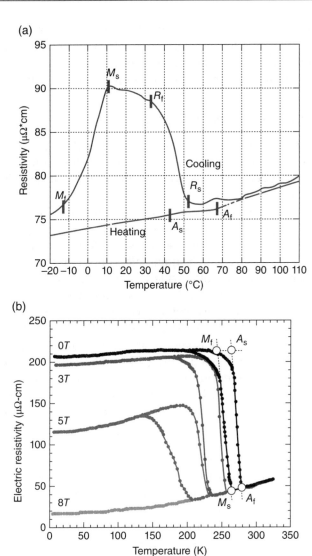

Figure 7.9 Electric resistivity response of NiTi (a) and NiMnCoIn alloys (b). Reproduced with permission from Ref. [31], Elsevier. Reproduced with permission from Ref. [32], the American Institute of Physics

7.3 Mechanical Characterization

Thermal analysis, magnetization, and resistivity experiments supply useful information about transformation temperatures, but still, they do not provide any information about the shape memory properties such as transformation strain and critical stress for transformation which are crucial parameters for actuator applications. Thus, thermomechanical experiments are required to quantify shape change in terms of strain during phase transformation.

7.3.1 Thermomechanical Testing

Thermal cycling under constant stress and stress cycling at a fixed temperature experiments are conducted to determine the transformation temperatures with respect to applied stress, recoverable and irrecoverable strain, temperature, and mechanical hysteresis.

The simplest strain measurements can be done through the measurement of the crosshead displacement of the testing machine or using extensometers on the samples. Strain gauges can also be used, but they are difficult to attach and are limited in measuring high strains. Noncontact measurements by video and laser extensometers are also widely used to measure the strain. Laser extensometers detect the reflected laser signal from the tags, which are fixed on the specimen, and obtain an accurate measurement of the distance between the edges of the tags. Overall, their strain accuracy is close to that of a mechanical extensometers. Video extensometers are very useful for testing very compliant specimens. A high-resolution digital camera takes the images of the specimen, while a computer processes the images in real time, noting the distance between two or more visible markers. Generally, video extensometers are not used for stiff metallic specimens where the strains are relatively small, but more advanced models are now available that provide sufficient accuracy at small strains [33].

Digital image correlation (DIC) is another noncontact optical method that can be used to measure displacements on the surface of an object, from which Lagrangian strains can be calculated [34]. To calculate full-field displacements, random markers are first applied to the surface of the test sample and then tracked in subsets as the sample is deformed. Surface deformation is calculated by comparing the deformed marker positions to their initial positions in the reference image. This is a powerful technique that gives the full-field strain distribution across a specimen, unlike conventional extensometer that gives a strain averaged over a single gauge length. The technique is fast, robust, and scalable and provides an accurate method to determine the surface geometry, displacements, and strains of a deforming object [35]. It is also possible to conduct DIC measurements in scanning electron microscope (SEM) to establish the relationship between localized strain and microstructure. For 3D DIC measurements, two or more cameras should be utilized to

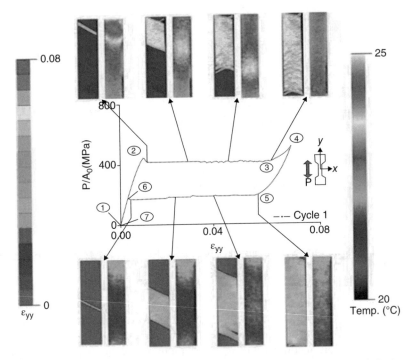

Figure 7.10 Macroscopic stress–strain curves with full-field maps of strain and temperature. Reproduced with permission from Ref. [35], Springer

capture the deformation of the samples from multiple viewpoints. Figure 7.10 shows the macroscopic stress–strain curves of a NiTi sample during a super-elasticity test, with combined images of the strain determined by DIC method and temperature determined by an infrared camera [35]. It is clear that DIC method could provide very useful information on the evolution of phase transformation during the test.

7.3.1.1 Shape Memory Effect

A schematic of the shape memory effect during the thermal cycling under constant stress is shown in Figure 7.11. Stress is applied in austenite and kept constant during the thermal cycling. Austenite transforms to martensite and stress-favored martensite variants are formed during cooling, resulting in a shape change. Upon heating, the shape change is recovered by martensite to austenite transformation. From these experiments, transformation tempera-tures, recoverable strain, and temperature hysteresis can be determined as a function of stress. It should be noted that, in general, shape change cannot

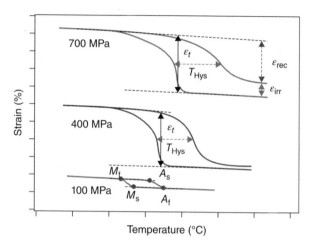

Temperature (°C)

Figure 7.11   Shape memory effect of $Ni_{50.4}Ti_{29.9}Hf_{19.3}Zr_{0.4}$ SMA as a function of applied stress

be observed if external stress is not applied due to the formation of self-accommodating (temperature-induced) martensite structure. Upon application of stress, self-accommodating martensite variants can be disrupted, and favored martensite variants can be formed, resulting in a shape change. The volume fraction of favored martensite variants and the transformation strain increase with stress. Figure 7.11 also covers the graphical method to determine transformation strain, temperatures, and hysteresis. If there is no plastic deformation, the recoverable strain is equal to the transformation and total strain. However, further increase in applied stress could result in plastic deformation. In this case, the total strain will be is the sum of recoverable and irrecoverable strains. It should also be noted that irrecoverable strain could stem from the plastic deformation and/or remnant martensite.

The transformation temperatures can be determined by the intersection method similar to the one applied to DSC tests. Total strain can be obtained from the difference between the strains of austenite and martensite at a selected temperature, as shown in Figure 7.11. Since austenite and martensite have different thermal expansion coefficients, total strain can be temperature dependent. Thus, in general, total strain is calculated at M_s and A_f or at the midtemperature of the thermal hysteresis.

Irrecoverable strain can be calculated at a temperature above A_f, and recoverable strain can be calculated either at the same temperature or by subtracting irrecoverable strain from total strain. There are several methods to determine temperature hysteresis: (i) the temperature difference during cooling and heating at the midpoint of the total strain, (ii) $(A_f - M_s)/2$, and (iii) $(A_f - M_f)/2$.

Figure 7.12 illustrates the transformation (or total) strain, irrecoverable strain, and temperature hysteresis of aged NiTiHfPd polycrystal [36] and NiTiHf single crystal [37] which are extracted from the thermal cycling under constant stress experiments. When the applied stress is not sufficient to induce favored martensite variants, mostly self-accommodated martensite variants form, resulting in low transformation strain. Transformation strain increases and then saturates with stress due to the increased and then saturated volume fraction of favored martensite variants. It should be noted that defect generation could occur as stress is increased, and as a consequence, irrecoverable strain also increases. It is also clear from Figure 7.12b that the shape memory characteristics (total strain, thermal hysteresis, and irrecoverable strain) of SMAs are highly orientation dependent. Total compressive strain does not change significantly with applied stress and is less than 1% along the [001] orientation, while it increased to 3% under 500 MPa along the [011] orientation of NiTiHf SMAs.

Thermal hysteresis is related to the energy dissipation during the phase transformation. Friction between the austenite and martensite interphases, dissipation during detwinning, incompatibility of transforming phases, and slip are the main factors of energy dissipation. In the absence of plastic deformation, hysteresis could either increase or decrease with stress, depending on the compatibility of the transforming phases. As shown in Figure 7.11, thermal hysteresis initially decreases with stress and then increases at higher stress levels. Similarly, thermal hysteresis decreases with stress in [001]-oriented NiTiHf SMAs where no plastic deformation is observed as shown in Figure 7.12b. The decrease in thermal hysteresis can be attributed to the increased volume fraction of favored martensite variants, which eliminates the dissipation energy between martensite variants. In all cases, thermal hysteresis increases with stress when plastic deformation is observed.

Work output of SMAs can be calculated by multiplying the recoverable strain and applied constant stress which can be determined from thermal cycling under constant stress experiments. Figure 7.13 shows the comparison of work output of NiTi-based shape memory alloys as a function of operating temperature [38]. It is clear that composition is an effective method to tailor the shape memory properties.

7.3.1.2 Superelasticity

Figure 7.14 shows the typical superelasticity behavior of shape memory alloys. The sample is loaded above A_f where elastic deformation of austenite is initially

(a)

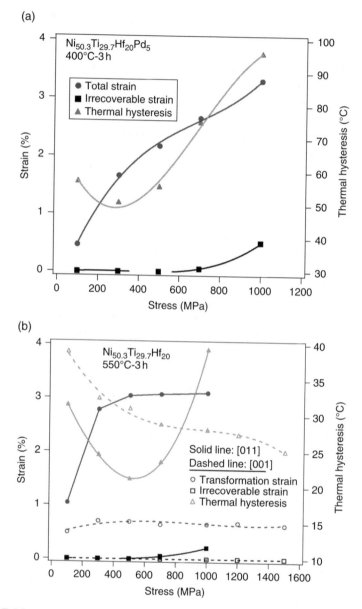

(b)

Figure 7.12 Total strain, irrecoverable strain, and thermal hysteresis of 400°C for 3-h aged NiTiHfPd polycrystalline alloy (a) and 550°C for 3-h aged [001] and [110] oriented NiTiHf single crystals (b) as a function of compressive stress

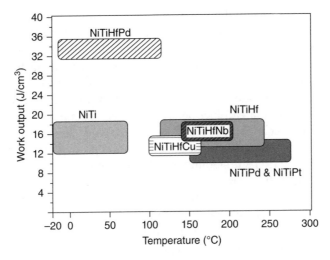

Figure 7.13 Work output comparison of NiTi-based shape memory alloys

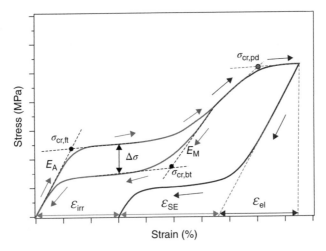

Figure 7.14 A schematic of the superelasticity behavior of shape memory alloys

observed. When the load reaches a critical value, stress-induced austenite to martensite transformation occurs where in most cases a plateau is observed at this stage. Further loading results in elastic deformation of martensite and detwinning of martensite. If the sample is unloaded at this region, shape recovery starts with elastic deformation recovery of the martensite, followed by the martensite to austenite transformation and elastic deformation recovery of austenite. If there is no plastic deformation, the fully reversible shape recovery will be obtained. On the other hand, if at the end of the austenite to martrensite transformation plateau the sample is further loaded, after (the second) elastic deformation of martensite the stress exceeds the critical stress for slip deformation where another plateau-like behavior can be observed. Upon unloading after this stage, specimen does not show full recovery due to plastic deformation. This plastic strain is due to the generation and propagation of dislocations which results in martensite stabilization. Young's moduli of austenite (E_A) and martensite (E_M), critical stresses of forward ($\sigma_{cr,ft}$) and backward ($\sigma_{cr,bt}$) transformations, critical stress for plastic deformation ($\sigma_{cr,pd}$), mechanical hysteresis (Δ_σ), elastic strain (ε_{el}), superelastic strain (ε_{SE}), and irrecoverable strain (ε_{irr}) can be determined from this type of isothermal mechanical testing.

Critical stresses for martensite reorientation, martensitic transformation, and slip are functions of the testing temperature. If the material is in martensite and deformed below A_s, as shown in Figure 7.15a, critical stress for the martensite reorientation decreases with temperature due to increased mobility of internal twins and martensite plates boundaries. If the material is in austenite and deformed between M_s and A_f ($M_s < T < A_f$), stress-induced martensite is formed during loading where critical stress for martensitic transformation increases with temperature and shape recovery cannot be observed upon unloading. Shape recovery occurs when the temperature is increased above A_f. Superelasticity is observed when the sample is deformed between A_f and M_d ($A_f < T < M_d$). Stress-induced martensite cannot be observed above M_d (martensite dead/desist temperature), and alloys deform like conventional materials [39]. M_d can be considered as the intersection of critical stresses of martensitic transformation and slip, as shown in Figure 7.15b. If the material is not strong enough or the testing temperature is close to M_d, partial recovery can be observed since martensitic transformation and plastic deformation occur simultaneously. If the sample is deformed above M_d, plastic deformation takes place before martensitic transformation where shape recovery cannot be observed during unloading. Thus, superelasticity can only be observed between A_f and M_d where the difference between these temperatures is called superelastic window (see Figure 7.15b).

Phase Diagram

Thermal cycling under constant stress experiments can be conducted in order to determine the transformation temperatures $\left(M_s^\sigma, M_f^\sigma, A_s^\sigma, \text{and } A_f^\sigma\right)$ as a function of stress by the tangent method, as shown in Figure 7.11. Moreover, critical stresses of forward and backward transformations

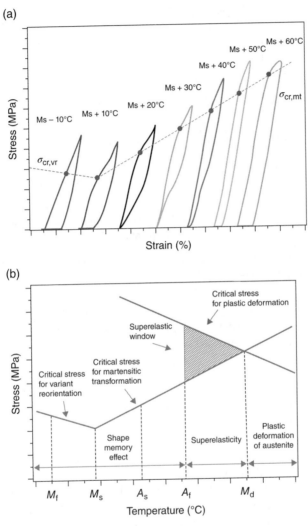

Figure 7.15 Schematic of stress–strain curves showing temperature dependency of the critical stress (a); relationship of critical stress and transformation temperatures (b) of SMAs

$(\sigma_{M_s}, \sigma_{M_f}, \sigma_{A_s},$ and $\sigma_{A_f})$ can be obtained from superelastic stress–strain curves as a function of temperature. These measured temperatures and stresses can be plotted to form the phase diagram of the shape memory alloy, as shown in Figure 7.16. In general, it is well known that in shape memory alloys, transformation temperatures increase linearly with stress and critical stresses

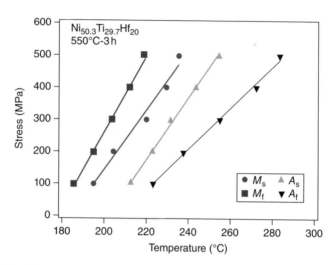

Figure 7.16   Phase diagram of aged NiTiHf shape memory alloys. Reproduced with permission from Ref. [41], Elsevier

increase linearly with temperature. The Clausius–Clapeyron (C–C) relationship can be used to predict the slope of the linear curve of M_s^σ and σ_{M_s}. The C–C equation can be expressed as

$$\frac{\Delta\sigma}{\Delta T} = \frac{\Delta H}{T_0\varepsilon_{tr}} \tag{7.8}$$

where ΔH is the changes in transformation enthalpy, T_0 is the equilibrium temperature, and ε_{tr} is the transformation strain. The C–C slope for equiatomic NiTi polycrystalline alloys is about 12 MPa/°C in compression [40] and 5–8 MPa/°C in tension [41]. The same constant is about 8 MPa/°C in tension [42] and 7–15 MPa/°C in compression [42, 43] for $Ni_{50.3}Ti_{29.7}Hf_{20}$. It is also worth noting that the C–C slopes are in the high range of 9–18 and 18–24 MPa/°C for high strength $Ni_{45.3}Ti_{29.7}Hf_{20}Pd_5$ [36] and $Ni_{45.3}Ti_{29.7}Hf_{20}Cu_5$ [44] polycrystalline alloys under compression, respectively.

Stress-State Effect

Mechanical behavior of NiTi shape memory alloys is stress state dependent. It means that the shape memory behavior can be different in tension, compression, and torsion. Tension testing methods and parameters of SMAs are identified in ASTM E8, "Standard test methods for tension testing of metallic materials." Figure 7.17 shows the tension, shear, and compression behaviors of NiTi. It is clear that the flat stress plateau (Lüders-like deformation)

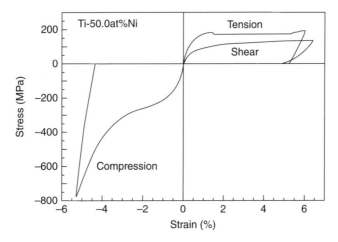

Figure 7.17   Deformation behavior of NiTi shape memory alloys under tensile, compressive, and shear stress. Reproduced with permission from Ref. [45], Elsevier

is observed in tension and torsion (shear) but not in compression. The asymmetry in the mechanical behavior of SMAs is related to the difference in deformation mechanisms during phase transformation and the martensite morphology. As an example, it is known that variant–variant interaction is more pronounced, and detwinning is more difficult under compression, resulting in higher stress–strain slope during transformation and lower transformation strain. Furthermore, observed twinning types can also be different.

Shape memory properties of SMAs are also strongly orientation dependent. Figure 7.18a shows the compressive response of an aged $Ni_{50.8}Ti_{49.2}$ (at.%) along the [001], [110], [111], and [148] orientations [46]. It is clear that [111]- and [112]-oriented single crystals exhibit high stress–strain slope during the martensitic transformation due to the formation of multiple correspondent variant pairs, while [001]- and [148]-oriented single crystals exhibit plateau-like behavior. The experimentally observed strain along the [148] orientation is very close to the theoretical transformation strain due to single corresponding variant pair (CVP) formation. Plastic deformation is avoided in [001] orientation due to zero Schmid factor for the two slip systems of {011} <100> and {001}<100> in NiTi. Figure 7.18b shows the shape memory effect of NiTiHf single crystals along the [011], [111], and [001] orientations under 500 MPa. It is clear that the thermally activated shape memory effect under constant stress is also highly orientation dependent [47, 48].

Stress hysteresis of shape memory alloys can be stress state and strain dependent. Lüders-like deformation propagates the stress-induced martensite

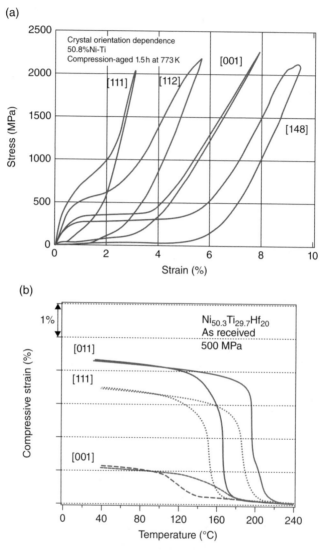

Figure 7.18 (a) Effects of orientation on the superelastic behavior of NiTi (Reproduced with permission from Ref. [46], Elsevier); (b) shape memory effect of NiTiHf

as localized deformation band, and transformation occurs at the same stress level within the strain limit. Deformation interfaces are recovered during the backward transformation, and stress hysteresis does not change with strain,

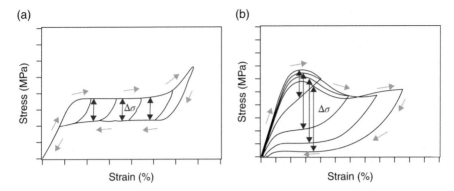

Figure 7.19 Mechanical hysteresis of shape memory alloys, strain dependent (a) or independent (b), based on the deformation mechanism

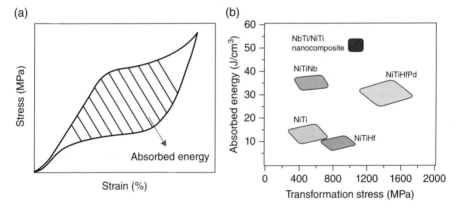

Figure 7.20 Damping measurements from superelastic loop (a), damping capacity of NiTi-based shape memory alloys (b) [36]

as shown in Figure 7.19a. If there is high interaction between the martensite plates or material is not strong enough, the dissipation energy increases with strain and results in increased mechanical hysteresis as a function of strain, as shown Figure 7.19b.

High dissipation due to the movement of phase front, martensite variant interfaces, and twin boundaries increases the mechanical hysteresis and damping capacity of shape memory alloys. Damping capacity of SMAs can be measured by the area enclosed by the superelastic loop, as shown in Figure 7.20a where high stress hysteresis and strain increase the damping capability. Damping capacity of NiTi-based alloys is shown in Figure 7.20b [36].

In order to achieve stable shape memory response and study the fatigue properties, thermal or mechanical cycling is conducted in shape memory alloys. For industrial applications, shape memory material is thermally cycled under constant stress or stress cycled at a selected temperature until the shape memory behavior stabilized, as shown in Figure 7.21. Created dislocations and retained martensite hinder further plastic deformation and dissipation with following cycles.

Thermal cycling with or without applied stress and isothermal superelastic cycling are the most efficient training methods to obtain two-way shape memory effect (TWSME). TWSME is a unique behavior of shape memory alloys where the material is capable of demonstrating reversible shape change during thermal cycling in the absence of external stress. TWSME is an interesting property for actuator applications since shape change occurs simply by changing the temperature. Dislocation generation and retained martensite are the main reasons for the creation of internal stress field that results in the formation of favored martensite variants during phase transformation. A TWSME response of $Ni_{45.3}Ti_{29.7}Hf_{20}Cu_5$ polycrystalline alloys is shown in Figure 7.22. There was no shape change before training due to the insufficient internal stress levels to disrupt self-accommodated martensite variants. After the training, a relatively small recoverable strain of 0.8% was observed by thermal cycling.

In shape memory alloys, martensitic transformation can occur in multiple stages depending on orientation, composition, stress state, and temperature. In NiTi, multiple-stage transformation exhibits transformation from austenite to R-phase and then to martensite. When Cu (near 10%) is added to NiTi, two-stage transformation from cubic austenite (B2) to orthorhombic martensite (B19) and then to monoclinic martensite (B19′) can be observed as shown in Figure 7.23a [50]. Multistage transformation can also be detected during pseudoelasticity experiments as shown in Figure 7.23b where a single crystalline of NiFeGa transforms from $L2_1$ austenite to a mixture of 10M and 14M modulated martensite to nonmodulated $L1_0$ martensite during loading [51].

Effects of Aging

Aging is one of the most efficient methods to tailor the shape memory properties. Heat treatments could result in precipitation formation in shape memory alloys that are essential to control the transformation temperatures and to increase material strength. Transformation temperatures of Ni-rich NiTi alloys are very sensitive to the Ni content of the matrix. These temperatures decrease about 93°C/at.% [52]. In Ni-rich NiTi alloys, formation of Ni_4Ti_3, Ni_3Ti_2, and Ni_3Ti precipitates deplete the Ni content of the matrix and thus

(a)

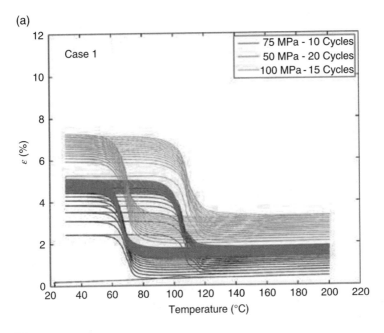

(b)

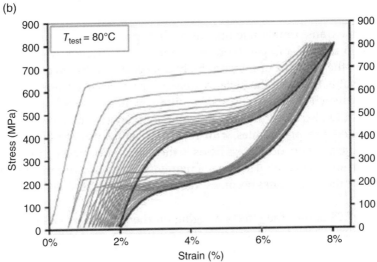

Figure 7.21 Thermal cycling under constant stress in SMAs (a) [49]; isothermal superelastic cycling in SMAs (b). Reproduced with permission from Ref. [46], Elsevier

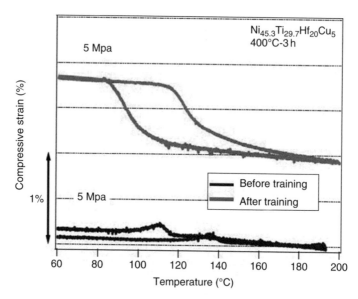

Figure 7.22 Two-way shape memory effect in NiTiHfCu polycrystalline alloys [44]

increase the transformation temperatures. The precipitates act as obstacles to dislocation motion and can therefore improve the strength of the material. However, the strength of material highly depends on the precipitation properties, that is, size and the distance between particles. Additionally, cooling procedures such as water quenching, oil quenching, air cooling, or furnace cooling could also affect the size and volume fraction of precipitates. Formation of nano-size precipitates with relatively small distance between them could create a strong local stress fields in the matrix surrounding the particles, which can suppress the nucleation of martensite. The increase in interparticle distance as the precipitates become larger in size diminishes the strength of the material.

Figure 7.24 shows the effects of aging on the DSC responses of a Ni-rich NiTiHf alloy. The transformation temperatures were initially decreased and then increased with aging temperature for the aging duration of 3 h (Figure 7.24a). For aging at 400°C, transformation temperatures were decreased with increasing aging time (Figure 7.24b). When the specimens aged at 500°C, transformation temperatures were slightly decreased and then increased with aging time (Figure 7.24c). It is clear from DSC responses that TTs initially decreased at low temperature or after aging for a short duration. This could be attributed to the existence of strong internal stress field due to

(a)

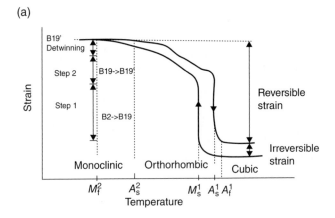

(b)

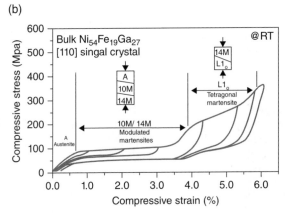

Figure 7.23 Two-stage martensitic phase transformation in NiTiCu (a) and NiFeGa (b) shape memory alloys. Reproduced with permission from Refs. [50, 51], Elsevier

formation of fine nanosize precipitates with small interparticle distance in the matrix which make the nucleation of martensite difficult; hence, more undercooling is needed for forward transformation to occur. However, heat treatments at high temperatures or for longer durations increased the size and/or volume fraction of precipitates, and as a result, more Ni content was depleted from the matrix, and TTs were shifted to higher temperatures. Thus, it is clear that both heat treatment temperature and duration are important parameters to control transformation temperatures.

Microstructure analysis of aged Ni-rich NiTiHf alloy is conducted by TEM and shown in Figure 7.25. Wide martensite plates with internal twins were observed in the as-extruded sample with no precipitates (Figure 7.25a). Aging

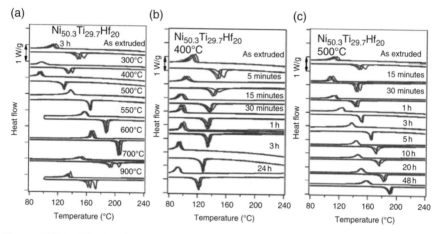

Figure 7.24 Effects of aging on the transformation temperatures of NiTiHf alloys for aging for 3 h (a), aging at 400°C (b), and aging at 550°C (c). Reproduced with permission from Ref. [43], Elsevier

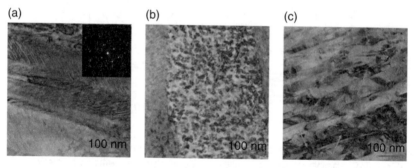

Figure 7.25 TEM microstructure analysis of NiTiHf shape memory alloys as-extruded condition (a), aged at 550°C for 3 h (b) and aged at 650°C for 3 h (c). Reproduced with permission from Ref. [43], Elsevier

at 550°C for 3 h has produced nano-size precipitates (20 nm) which is uniformly distributed in the matrix (Figure 7.25b). Precipitate size was increased around 40–60 nm with increasing aging temperature to 650°C (Figure 7.25c). Interparticle distance was increased as a result of increasing precipitate size. It is worth noting that the morphology of martensite phase is highly related to the size and volume fraction of precipitates.

Effects of aging on the shape memory effect and superelastic behavior of Ni-rich NiTiHf alloys are shown in Figure 7.26. It is clear that transformation

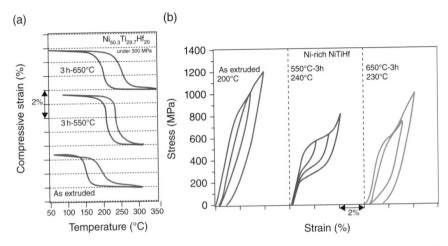

Figure 7.26 Aging effect on shape memory effect (a) and superelastic behavior (b) in NiTiHf shape memory alloys. Reproduced with permission from Ref. [43], Maney Publishing

temperatures, thermal hysteresis, and transformation strain can be tailored by aging. Superelastic behavior was improved for the sample aged at 550°C for 3 h due to formation of coherent nano-size precipitates. It is evident from Figure 7.26a that transformation strain was increased above 3% after aging at 550°C/3 h which can be attributed to absorption of precipitates by martensite plates during forward transformation as shown in Figure 7.25 (TEM images). However, precipitates became larger in size after aging at 650°C/3 h, and growth of martensite variants was limited in the space between precipitates.

Figure 7.27 shows the cyclic DSC responses of the solutionized and 550°C for 3-h aged Ni-rich NiTiHf shape memory alloys. Transformation temperatures of solutionized samples were shifting to lower temperatures with increasing number of cycles due to low strength of material. Aging at 500°C for 3 h improves the thermal stability where transformation temperatures were almost remained constant.

Shape memory properties and martensite morphology depend on the precipitate characteristics (size and density of precipitates and interparticle distance). Aging at low temperatures/short time result in the formation of fine nanometer size precipitates with small interparticle distance which enhances the strength of the SMAs. Therefore, shape memory properties and cyclic stability were improved. However, size of precipitates and the distance between them increase after aging at high temperatures/long durations,

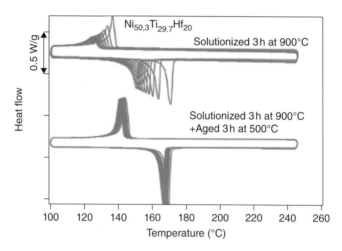

Figure 7.27   Aging effect on cycling response of NiTiHf alloy. Reproduced with permission from Ref. [43], Elsevier

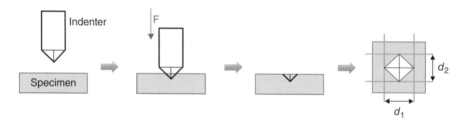

Figure 7.28   Schematic of Vickers hardness measurement

which diminish the strength of material. In addition, martensite morphology (e.g., twinning type and thickness) can also be modified. Thus, it is reasonable to conclude that thermal treatment is a very efficient tool to alter the microstructure and shape memory properties.

Hardness and Indentation

Hardness is the property of a material that measures its resistance to plastic deformation by penetration. Figure 7.28 shows that an indenter is pressed into the surface of the metal to be tested under a specific load for a definite time interval, and a measurement is made of the size or depth of the indentation. There are three principal standard test methods for expressing the relationship between hardness and the size of the impression: Brinell, Rockwell, and Vickers. In the Vickers hardness test, accurate readings can be taken, and just

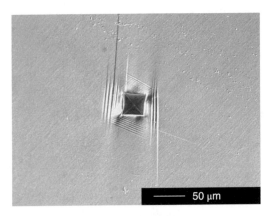

Figure 7.29 Indentation-induced martensitic transformation adjacent indent

one type of indenter is used for all types of metals and surface treatments. The theory and practice of the test is described in ASTM E92—"Standard test method for Vickers hardness of metallic materials." The two diagonals of the indentation left in the surface of the material are measured using a microscope, and the area of the sloping surface of the indentation is calculated.

In shape memory alloys, hardness of the transforming phases could be different. In NiTi, austenite has higher hardness than martensite. Moreover, depending on the testing temperature, stress-induced austenite to martensite transformation or variant reorientation could occur during the hardness test. Figure 7.29 shows the optical image of a CoNiAl alloy where hardness test results in austenite to martensite transformation. It is clear that stress-induced martensite variants were formed after indentation. Thus, one should be careful to interpret and compare the hardness results.

Indentation Method

Indentation is a technique to characterize the local shape memory properties of SMAs in nano and microscales. Nominal stress–strain curve, elastic modulus, as well as hardness can be obtained from spherical indentation with satisfactory precision [53–55]. Indenter tip (Berkovich, spherical, etc.) and size can be selected according to the application. Specimens should be mechanically polished before indentation to obtain accurate results. From experimental readings of indenter, load and depth of penetration elastic modulus and hardness of the specimen can be extracted. In a typical test, force and depth of penetration are recorded as load is applied from zero to a maximum and then from maximum force back to zero. Figure 7.30 is a schematic

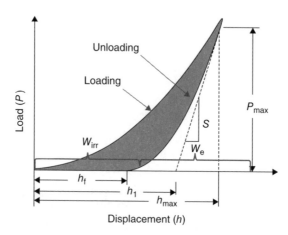

Figure 7.30 Schematic loading-unloading curve from nanoindentation experiment

representation of an indentation load–displacement data where P_{max} is the maximum indentation load, h_{max} is the maximum indenter displacement at peak load, h_f is the final depth of the contact impression after unloading, and S is dP/dh. Since the size of impression is small, instead of optical instruments, depth of penetration and geometry of indent are measured in the area of contact. The analysis of the initial portion of this elastic unloading response gives an estimate of the elastic modulus of the indented material.

The extent of the shape recovery of SMAs is determined by the material, temperature, and indentation load and shape of the indenter. For instance, complete recovery can be observed during the spherical indentation of superelastic NiTi alloys under low loads [56, 57]. The total work W_t required to move an indenter into a solid is the area under the loading curve, and the reversible work W_e is the area under the unloading curve and W_{irr} is the irreversible work. The recovery ratio is defined as the ratio of reversible work to the total work [58]:

$$\eta_w = \frac{w_e}{w_t} = \frac{\int_{h_r}^{h_{max}} Fdh}{\int^{h_{max}} Fdh} \tag{7.9}$$

Figure 7.31 shows the indentation curves of superelastic NiTi and copper with spherical and Berkovich indenters. It is clear that indentation response

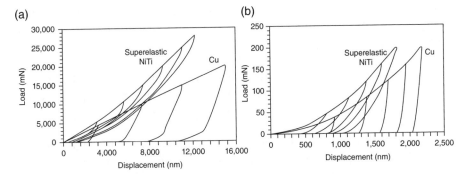

Figure 7.31 Indentation results of NiTi and copper by using spherical indenter (a), Berkovich indenter (b). Reproduced with permission from Ref. [58], the American Institute of Physics

of NiTi is highly recoverable when it is compared to copper. The work recovery ratios of NiTi are 90% for the spherical indenter and 45% for the Berkovich indenter which are substantially higher than the values of copper [58].

7.4 Microstructural Analysis

Shape memory properties are highly microstructure dependent, so it is crucial to determine the microstructural features of SMAs. The microstructure of the material depends on the fabrication methods, composition, and thermomechanical treatments.

The microstructural features such as grain size, orientation of grains, precipitates, inclusions, impurities, second phases, porosity, segregation, or surface effects can be determined by using optical or electron microscopes. Specimen preparation is an important part of metallography and consists of sample selection, sectioning, grinding, polishing, and etching. There are several techniques to characterize the microstructure of materials including optical microscopy, SEM and transmission electron microscopy (TEM).

7.4.1 Optical Microscopy

Optical microscopy can be used to determine the martensite morphology, grain size, and second-phase formation. Sample preparation may change the microstructure of the material. As an example, cutting with abrasives may cause a large amount of damage. After cutting, the specimen needs to be mounted for polishing. It should be noted that some mounting techniques could change the temperature of the sample. In general, epoxy resin and hardeners are used for mounting. The grinding procedure involves several stages, using a finer paper (higher number) for each successive stage. Each grinding stage removes the scratches from the previous coarser paper, and the specimen should be washed thoroughly with soapy water to prevent the contamination from coarse grit. For polishing, diamond suspensions of 9, 6 and 3 µm are used. Finally, alumina suspensions of 1 and 0.5 µm are used to produce a smooth surface.

Etching is used to reveal the microstructure of the metal through selective chemical solution. It also removes the thin, highly deformed layer introduced during grinding and polishing. The etching solution H_2O (82.7%) + HNO_3 (14.1%) + HF (3.2%) is generally used for NiTi, while HCl (75 ml) + ethanol (75 ml) + $CuSO_4$ (15 g) + distilled water (10 ml) is used for magnetic shape memory alloys.

Figure 7.32 shows the optical microscopy images of NiTi and CoNiAl after etching. Figure 7.32a reveals the formation of Ni_3Ti in NiTi alloys. Figure 7.32b can be used to determine the grain size and martensite variants in equiatomic NiTi alloys. Figure 7.32c shows the martensite morphology of CoNiAl alloys.

One of the limitations of the optical microscope is that of resolution. High-resolution imaging is more commonly carried out in a SEM. Figure 7.33 depicts a SEM micrograph of Ti 50.8 at.%Ni that shows the formation of a self-accommodating martensite variant [59].

7.4.2 XRD

XRD can be used to determine the crystal structure (crystalline state, interatomic distance, and bond angle) and the lattice parameters of the phases present in the microstructure. In XRD, a detector moves in a circle around the sample, and its position and number of X-rays observed are recorded as the angle (2θ). Then, Bragg's law is used to calculate the inter-plane spacing, d:

$$\lambda = 2d \sin \theta \qquad (7.10)$$

where λ is the wavelength and θ is the angle. Next, spacing between the planes can be calculated by using the Miller indices. As an example, the interplanar distance of a cubic structure can be found as

$$d_{hkl} = \frac{a}{\sqrt{h^2 + k^2 + l^2}} \qquad (7.11)$$

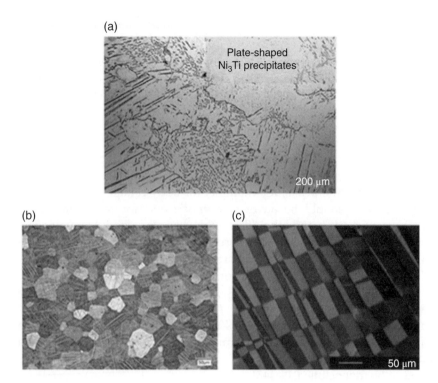

Figure 7.32 Optical images of SMAs where Ni_3Ti precipitates formation in NiTi (a) grain size of NiTi (b) and martensite morphology of CoNiAl (c) are revealed

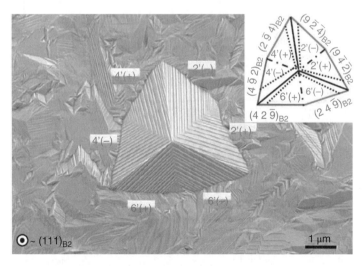

Figure 7.33 SEM micrograph of the reverse transformation relief showing hexangular morphology of self-accommodated martensite [59]

where a is the lattice parameter. It should be noted that interplanar distance equation is different for each crystal structure. Equation 7.10 can also be written as

$$\frac{\lambda^2}{4a^2} = \frac{\sin^2\theta}{h^2 + k^2 + l^2} \qquad (7.12)$$

If the angle of the diffraction peak and the Miller indices of the corresponding plane are known, the lattice parameter can be calculated. There are certain rules to observe the peaks, for example, for body-centered cubic structures (BCC), $h + k + l$ should be an even number, and for face-centered cubic (FCC) structures, h, k, and l should be either all even or odd. Manual indexing is time-consuming but still very useful. Autoindexing is done by using computer-based indexing software products.

Figure 7.34 illustrates the results of the XRD scans carried out at 25 and 200°C on the as-extruded $Ni_{50.3}Ti_{29.7}Hf_{20}$ alloy. The as-extruded sample was in the martensitic state at 25°C and fully austenitic at 200°C. At 200°C, a major peak was observed at about 42°, and the lattice structure of austenite was determined to be B2. At room temperature, major XRD peaks of martensite were observed between 35° and 50°. The martensite phase for the alloy was found to be monoclinic B19' [43].

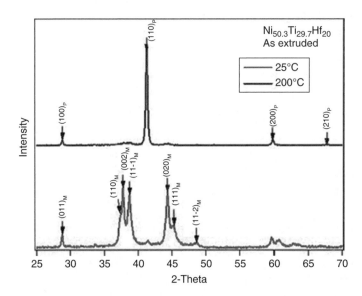

Figure 7.34   XRD results for the as-extruded $Ni_{50.3}Ti_{29.7}Hf_{20}$ alloy. Reproduced with permission from Ref. [43], Elsevier

It is noteworthy that conventional XRD can also be used to determine the texture as well as internal stress of SMAs, but the information can only be obtained from the surface of the samples. On the other hand, advanced diffraction techniques such as neutron and synchrotron diffraction can provide information on the bulk samples due to their ability to penetrate much deeper. *In situ* neutron and synchrotron diffraction studies could provide vital information on texture, strain, deformation modes (e.g., slip and twinning), and phase fraction evolution with changes in temperature and stress.

7.5 Summary

Common experimental characterization methods to determine the shape memory properties were explained in this chapter. Transformation temperatures, recoverable strain, hysteresis, and other important parameters can be determined by several methods. Experimental characterization of shape memory behavior and microstructure are crucial to determine the shape memory properties for modeling and application purposes.

References

[1] P. Wollants, J. Roos, L. Delaey, Progress in Materials Science, **37** (1993) 227–288.

[2] E. Panchenko, Y. Chumlyakov, I. Kireeva, A. Ovsyannikov, H. Sehitoglu, I. Karaman, Y. Maier, The Physics of Metals and Metallography, **106** (2008) 577–589.

[3] K. Otsuka, X. Ren, Progress in Materials Science, **50** (2005) 511–678.

[4] H. Sehitoglu, C. Efstathiou, H.J. Maier, Y. Chumlyakov, Mechanics of Materials, **38** (2006) 538–550.

[5] H. Sehitoglu, I. Karaman, X.Y. Zhang, Y. Chumlyakov, H.J. Maier, Scripta Materialia, **44** (2001) 779–784.

[6] R.F. Hamilton, H. Sehitoglu, Y. Chumlyakov, H.J. Maier, Acta Materialia, **52** (2004) 3383–3402.

[7] K. Kus, T. Breczko, Materials Physics and Mechanics, **9** (2010) 75–83.

[8] G. Della Gatta, M.J. Richardson, S.M. Sarge, S. Stølen, Pure and Applied Chemistry, **78** (2006) 1455–1476.

[9] D.C. Lagoudas, Shape Memory Alloys. Science and Business Media, LLC, New York, 2008.

[10] M. Kök, Z. Yakinci, A. Aydogdu, Y. Aydogdu, Journal of Thermal Analysis and Calorimetry, **115** (2014) 555–559.

[11] K. Nurveren, A. Akdoğan, W. Huang, Journal of Materials Processing Technology, **196** (2008) 129–134.

[12] C.A. Canbay, A. Aydoğdu, Journal of Thermal Analysis and Calorimetry, **113** (2013) 731–737.

[13] L. Kaufman M. Cohen, Progress in Metal Physics, **7** (1958) 165–246.

[14] R. Salzbrenner, M. Cohen, Acta Metallurgica, **27** (1979) 739–748.

[15] H.E. Kissinger, Analytical Chemistry, **29** (1957) 1702–1706.

[16] N. Mehta, P. Agarwal, A. Kumar, Turkish Journal Physics, **29**(3) (2005) 193–200.

[17] T. Ozawa, Journal of Thermal Analysis and Calorimetry, **2** (1970) 301–324.

[18] K. Gall, H. Sehitoglu, Y.I. Chumlyakov, Y.L. Zuev, I. Karaman, Scripta Materialia, **39** (1998) 699–705.

[19] Y. Liu, M. Blanc, G. Tan, J. Kim, S. Miyazaki, Materials Science and Engineering A, **438** (2006) 617–621.

[20] A. Khalil, A. Dlouhy, G. Eggeler, Acta Materialia, **50** (2002) 4255–4274.

[21] W.O. Soboyejo, T.S. Srivatsan, Advanced Structural Materials: Properties, Design Optimization, and Applications. CRC Press, Boca Raton, FL, 2006.

[22] D. Holec, Diploma Thesis, On the precipitation in NiTi based Shape Memory Alloys, Masaryk University, Brno, 2005.

[23] J. Khalil-Allafi, G. Eggeler, W.W. Schmahl, D. Sheptyakov, Materials Science and Engineering A, **438–440** (2006) 593–596.

[24] J. Khalil Allafi, X. Ren, G. Eggeler, Acta Materialia, **50** (2002) 793–803.

[25] L. Bataillard, J-E. Bidaux, R. Gotthardt, Philosophical Magazine A, **78** (1998) 327–344.

[26] M. Nishida, T. Hara, T. Ohba, K. Yamaguchi, K. Tanaka, K. Yamauchi, Materials Transactions, **44** (2003) 2631–2636.

[27] H. Karaca, I. Kaya, H. Tobe, B. Basaran, M. Nagasako, R. Kainuma, Y. Chumlyakov, Materials Science and Engineering A, **580** (2013) 66–70.

[28] Y. Kaynak, H.E. Karaca, I.S. Jawahir, Procedia CIRP, **13** (2014) 393–398.

[29] R. Zarnetta, R. Takahashi, M.L. Young, A. Savan, Y. Furuya, S. Thienhaus, B. Maaß, M. Rahim, J. Frenzel, H. Brunken, Advanced Functional Materials, **20** (2010) 1917–1923.

[30] H.E. Karaca, I. Karaman, B. Basaran, Y. Ren, Y.I. Chumlyakov, H.J. Maier, Advanced Functional Materials, **19** (2009) 983–998.

[31] V. Antonucci, G. Faiella, M. Giordano, F. Mennella, L. Nicolais, Thermochimica Acta, **462** (2007) 64–69.

[32] W. Ito, K. Ito, R.Y. Umetsu, R. Kainuma, K. Koyama, K. Watanabe, A. Fujita, K. Oikawa, K. Ishida, T. Kanomata, Applied Physics Letters, **92** (2008) 021908.

[33] C. Churchill, J. Shaw, M. Iadicola, Experimental Techniques, **33** (2009) 51–62.

[34] A. Kammers, S. Daly, Measurement Science and Technology, **22** (2011) 125501.

[35] K. Kim, S. Daly, Experimental Studies of Phase Transformation in Shape Memory Alloys, in: Mechanics of Time-Dependent Materials and Processes in Conventional and Multifunctional Materials, Volume 3. Springer, New York, 2011, pp. 81–87.

[36] H. Karaca, E. Acar, G. Ded, B. Basaran, H. Tobe, R. Noebe, G. Bigelow, Y. Chumlyakov, Acta Materialia, **61** (2013) 5036–5049.

[37] S.M. Saghaian, H.E. Karaca, H. Tobe, M. Souri, R. Noebe, Y.I. Chumlyakov, Acta Materialia, **87** (2015) 128–141.

[38] H. Karaca, E. Acar, H. Tobe, S. Saghaian, Materials Science and Technology, **30** (2014) 1530–1544.

[39] O. Benafan, R. Noebe, S. Padula II, A. Garg, B. Clausen, S. Vogel, R. Vaidyanathan, International Journal of Plasticity, **51** (2013) 103–121.

[40] L. Orgéas, D. Favier, Acta Materialia, **46** (1998) 5579–5591.

[41] Y. Liu, A. Mahmud, F. Kursawe, T-H. Nam, Journal of Alloys and Compounds, **449** (2008) 82–87.

[42] G.S. Bigelow, A. Garg, S.A. Padula Ii, D.J. Gaydosh, R.D. Noebe, Scripta Materialia, **64** (2011) 725–728.

[43] H. Karaca, S. Saghaian, G. Ded, H. Tobe, B. Basaran, H. Maier, R. Noebe, Y. Chumlyakov, Acta Materialia, **61** (2013) 7422–7431.

[44] H.E. Karaca, E. Acar, G.S. Ded, S.M. Saghaian, B. Basaran, H. Tobe, M. Kok, H.J. Maier, R.D. Noebe, Y.I. Chumlyakov, Materials Science and Engineering A, **627** (2015) 82-94.

[45] P. Sittner, Y. Liu, V. Novák, Journal of the Mechanics and Physics of Solids, **53** (2005) 1719–1746.

[46] J. Ma, I. Karaman, R.D. Noebe, International Materials Reviews, **55** (2010) 257–315.

[47] H.E. Karaca, S.M. Saghaian, B. Basaran, G.S. Bigelow, R.D. Noebe, Y.I. Chumlyakov, Scripta Materialia, **65** (2011) 577–580.

[48] A.P. Stebner, G.S. Bigelow, J. Yang, D.P. Shukla, S.M. Saghaian, R. Rogers, A. Garg, H.E. Karaca, Y. Chumlyakov, K. Bhattacharya, R.D. Noebe, Acta Materialia, **76** (2014) 40–53.

[49] S. Padula II, D. Gaydosh, A. Saleeb, B. Dhakal, Experimental Mechanics, **54** (2014) 709–715.

[50] H. Sehitoglu, I. Karaman, X. Zhang, A. Viswanath, Y. Chumlyakov, H. Maier, Acta Materialia, **49** (2001) 3621–3634.

[51] N. Ozdemir, I. Karaman, N. Mara, Y. Chumlyakov, H. Karaca, Acta Materialia, **60** (2012) 5670.

[52] W. Tang, Metallurgical and Materials Transactions A, **28** (1997) 537–544.

[53] L. Qian, S. Zhang, D. Li, Z. Zhou, Journal of Materials Research, **24** (2009) 1082–1086.

[54] M.F. Doerner, W.D. Nix, Journal of Materials Research, **1** (1986) 601–609.

[55] W.C. Oliver, G.M. Pharr, Journal of Materials Research, **7** (1992) 1564–1583.

[56] X. Fei, Y. Zhang, D.S. Grummon, Y-T. Cheng, Journal of Materials Research, **24** (2009) 823–830.

[57] Y. Zhang, Y-T. Cheng, D.S. Grummon, Applied Physics Letters, **89** (2006) 041912.

[58] W. Ni, Y.-T. Cheng, D.S. Grummon, Applied Physics Letters, **82** (2003) 2811–2813.

[59] M. Nishida, T. Nishiura, H. Kawano, T. Inamura, Philosophical Magazine, **92** (2012) 2215–2233.

Index

Shape Memory Alloy Actuators: Design, Fabrication, and Experimental Evaluation, First Edition. Mohammad H. Elahinia.
© 2016 John Wiley & Sons, Ltd. Published 2016 by John Wiley & Sons, Ltd.